The
Moment-SOS
Hierarchy

Lectures in Probability, Statistics, Computational
Geometry, Control and Nonlinear PDEs

Series on Optimization and its Applications

Print ISSN: 2399-1593
Online ISSN: 2399-1607

(*formerly known as Imperial College Press Optimization Series* —
Print ISSN: 2041-1677)

Series Editor: Jean Bernard Lasserre *(LAAS-CNRS and Institute of
Mathematics, University of Toulouse, France)*

Published

Series on Optimization and its Applications – Vol. 4

The Moment-SOS Hierarchy

Lectures in Probability, Statistics, Computational Geometry, Control and Nonlinear PDEs

Didier Henrion

LAAS-CNRS, Toulouse, France
University of Toulouse, France
Czech Technical University in Prague, Czech Republic

Milan Korda

LAAS-CNRS, Toulouse, France
University of Toulouse, France
Czech Technical University in Prague, Czech Republic

Jean B. Lasserre

LAAS-CNRS, Toulouse, France
Institute of Mathematics, University of Toulouse, France

W̶ World Scientific

NEW JERSEY · LONDON · SINGAPORE · BEIJING · SHANGHAI · HONG KONG · TAIPEI · CHENNAI · TOKYO

Published by

World Scientific Publishing Europe Ltd.

57 Shelton Street, Covent Garden, London WC2H 9HE

Head office: 5 Toh Tuck Link, Singapore 596224

USA office: 27 Warren Street, Suite 401-402, Hackensack, NJ 07601

Library of Congress Cataloging-in-Publication Data
Names: Henrion, D. (Didier), author. | Korda, Milan, author. |
   Lasserre, Jean-Bernard, 1953–   author.
Title: The moment-SOS hierarchy : lectures in probability, statistics, computational geometry,
   control and nonlinear PDEs / Didier Henrion, Milan Korda, Jean Bernard Lasserre.
Description: New Jersey : World Scientific, [2021] | Series: Series on optimization and
   its applications, 2399-1593 ; vol. 4 | Includes bibliographical references and index.
Identifiers: LCCN 2020030585 | ISBN 9781786348531 (hardcover) | ISBN 9781786348548 (ebook)
Subjects: LCSH: Mathematical optimization. | Moment problems (Mathematics) |
   Programming (Mathematics) | Optimal designs (Statistics) | Probabilities.
Classification: LCC QA402.5 .H456 2020 | DDC 519.6--dc23
LC record available at https://lccn.loc.gov/2020030585

British Library Cataloguing-in-Publication Data
A catalogue record for this book is available from the British Library.

For any available supplementary material, please visit
https://www.worldscientific.com/worldscibooks/10.1142/Q0252#t=suppl

Preface

In the last ten years the *Moment-Sums-Of-Squares Hierarchy* (in short the Moment-SOS Hierarchy) whose prototype was initially described in Lasserre (2000, 2001) for polynomial optimization, has emerged as a powerful technique to help solve difficult problems in many (various) branches of Mathematics with important real applications.

Why is this technique widely applicable (at least in principle)?

Because it applies to virtually *all* problems modeled as an instance of the so-called *Generalized Moment Problem* ("GMP" in the sequel) provided that the problem to solve can be described in terms of algebraic objects, namely, *polynomials* and *semi-algebraic sets*. (In fact even semi-algebraic functions are tolerated.) Indeed it has been known for a long time that the GMP has great modeling power with impact in several branches of Mathematics with a lot of important applications in various fields (see e.g. the collection of chapters in Landau (1987)). However in full generality the GMP is intractable and it was only in the early 2000's that one has realized how to use powerful results from Real Algebraic Geometry to provide a general technique to solve GMP with algebraic data.

In Kemperman (1987) the author pointed out the lack of a general algorithmic approach to solve the GMP. Indeed (quoting Kemperman, p. 20) "... a deep study of algorithms has been rare so far in the theory of moments, except for certain very specific practical applications, for instance, to crystallography, chemistry and tomography. No doubt, there is a considerable need for developing reasonably good numerical procedures for handling the great variety of moment problems which do arise in pure and applied mathematics and in the sciences in general...".

The Moment-SOS hierarchy fills up this gap for the class of GMP described by algebraic objects.

In short the Moment-SOS hierarchy works in three steps:

(1) It transforms the initial problem to solve into an infinite-dimensional Linear Programming (LP) problem $\mathbf{Q}$ on an appropriate space of (possibly signed) measures.

(2) Using the fact that the problem is described in terms of algebraic objects, $\mathbf{Q}$ is in turn converted into an equivalent infinite-dimensional *semidefinite program* (SDP) $\mathbf{P}$ on appropriate spaces of sequences (the sequences of moments of the measures involved in $\mathbf{Q}$). Problem $\mathbf{P}$ is an SDP because one uses powerful results on the $\mathbf{K}$-moment problem (where $\mathbf{K}$ is semi-algebraic) to provide necessary and sufficient conditions on real sequences to be *moments* of some (positive) measure on $\mathbf{K}$.

(3) Finally, the infinite-dimensional SDP $\mathbf{P}$ is in turn converted into a nested sequence $(\mathbf{P}_d)_{d\in\mathbb{N}}$ of finite-dimensional SDPs by finite truncation of moment sequences in $\mathbf{Q}$, up to order d. The associated monotone sequence of optimal values $(\rho_d)_{d\in\mathbb{N}}$ converges to the optimal value of $\mathbf{P}$ (hence of $\mathbf{Q}$). Moreover, depending on the problem on hand, optimal solutions of $(\mathbf{P}_d)$ provide some appropriate information on solutions of the initial problem.

Of course the no-free lunch rule applies and therefore this technique comes with a price to pay as in principle it can solve (or help solve) very difficult problems. Its computational cost can become high even for problems of modest dimension, especially in view of the state-of-the-art of semidefinite solvers that are used as a black-box in the general numerical scheme underlying the Moment-SOS hierarchy. Fortunately, very often, higher dimensional problems possess some inherent sparsity and/or symmetry in their description. This feature of large-scale problems is essential as sometimes it can be exploited successfully in the Moment-SOS hierarchy.

At this stage a remark is in order. As already mentioned, the Moment-SOS hierarchy relies on powerful results from Real Algebraic Geometry on a certain SOS-based representation of polynomials that are positive on a compact basic semi-algebraic set **K**. Such SOS-based representation of a polynomial provides its (algebraic) certificate of positivity on **K**. These results have a dual facet related to the **K**-moment problem which states necessary and sufficient conditions for a real sequence to be moments of a Borel measure supported on **K**. But there exist other positivity certificates that are not based on SOS and which can be obtained by solving LP or second-order cone programs instead of SDPs. So a more generic name for the approach would be: the *Moment-Positivity-Certificates hierarchy* as in principle there is no obligation to enforce a particular choice for the positivity certificates that are used. In practice however (at least for certain types of problems), the SOS-based certificates are the more efficient (but unfortunately also the most expensive).

The goal of these Lectures is to describe several new applications (not covered in Lasserre (2009)) of this powerful technique in various contexts, ranging from Computational Geometry, Probability, Statistics, Control and Optimal Control of Dynamical Systems governed by ordinary differential equations, Analysis and Control of Non-Linear Partial Differential Equations (Non-Linear PDEs), as well as some inverse problems like for instance the Inverse Optimal Control problem.

The book is divided into two main parts, and here is a chapter-by-chapter brief description of its content. After a chapter introducing the notation and some basic preliminary results used throughout the other chapters, Part I is devoted to application of the Moment-SOS hierarchy for solving problems in Probability, Statistics and Computational Geometry. Part II is devoted to application of the Moment-SOS hierarchy to solve problems in Control, Optimal Control of Dynamical Systems, as well as Analysis and Control of Non-Linear Partial Differential Equations.

Part I

Chapter 2 is concerned with a general overview of the basic principle behind the Moment-SOS hierarchy. Each of the following chapters is an illustration of this principle in various specific applications.

Chapter 3 is concerned with *"measuring semi-algebraic sets"*, that is, given some measure μ and some semi-algebraic set $\Omega \subset \mathbb{R}^n$ the goal is to evaluate $\mu(\Omega)$ based on the sole knowledge of the moments of μ.

Chapter 4 is concerned with computing the Lebesgue decomposition $\phi+\psi$ of a measure μ with respect to another measure λ (i.e. $\phi+\psi = \mu$ with $\phi \ll \lambda$ and $\psi \perp \lambda$), again based on the sole knowledge of the moments of μ and λ.

Chapter 5 is concerned with Super-Resolution.

Chapter 6 is concerned with sparse Polynomial interpolation which is shown to be a particular case of *discrete* Super-Resolution.

Chapter 7 is concerned with the representation of probabilistic constraints (chance-constraints).

Chapter 8 is concerned with the *Optimal Design* problem in Statistics.

Part II

In Part II we are concerned with several issues in Control and Optimal Control of Dynamical Systems, as well as the analysis and control of some Non-Linear Partial Differential Equations. A so-called *"weak formulation"* of such problems is the backbone and common feature of all chapters. It permits to introduce the central notion of *"occupation measures"* which is crucial to the application of the Moment-SOS hierarchy to solve the problem. One important and characteristic feature of this computational approach is to avoid discretizing the time interval (in optimal control) and the time-space domain (in PDEs). Instead the strategy is to compute moments of the measure supported on the trajectory of the solution.

Chapter 9 is concerned with the optimal control of dynamical systems.

Chapter 10 is concerned with the convex computation of region of attraction for dynamical systems (with and without control).

Chapter 11 is concerned with the analysis and control of some types of Non-Linear Partial Differential Equations.

Finally, **Chapter 12** is concerned with some additional material.

• At some places in the book, some theorems, lemmas or propositions are framed in a box to emphasize their importance, at least to the authors' taste.

• Sometimes, proofs of theorems, lemmas and propositions are postponed for the sake of clarity of exposition and so that the reader will not be lost in technical details in the middle of a chapter. Sometimes, if short and/or important, a proof is provided just after the corresponding theorem, lemma or proposition. Finally, if it is too long, or too technical, or not crucial, a proof is simply omitted but a reference is provided in the *Notes and sources* section at the end of the corresponding chapter.

Bibliography

Lasserre, J. B. (2000). *Optimisation globale et théorie des moments*, C. R. Acad. Sci. Paris **331**, Série 1, pp. 929–934.

Lasserre, J. B. (2001). *Global optimization with polynomials and the problem of moments*, SIAM J. Optim. **11**, pp. 796–817.

Kemperman, J. H. B. (1987). *Geometry of the Moment Problem*, In *Moments in Mathematics*, H. J. Landau (ed.), Proc. Symp. Appl. Math. **37**, American Mathematical Society, Providence, pp. 16–53.

Landau, H. J. (1987). *Moments in Mathematics*, Proc. Symp. Appl. Math. **37**, American Mathematical Society, Providence, USA.

Lasserre, J. B. (2009). *Moments, Positive Polynomials and Their Applications*, Imperial College Press, London, UK.

Acknowledgements

The research of Jean B. Lasserre and Didier Henrion was partially funded by the European Research Council (ERC) under the European Union's Horizon 2020 research and innovation program (grant agreement ERC-ADG 666981 TAMING).

Toulouse, June 2020   *Didier Henrion, Milan Korda, Jean B. Lasserre*

Contents

Chapter 1

Notation, Definitions and Preliminaries

We introduce the notation as well as some basic notions and preliminary results.

1.1 Notation and definitions

1.1.1 *Semidefinite programming*

Let $\mathcal{S}^m$ (resp. $\mathcal{S}^m_+$) be the convex cone of $m \times m$ real symmetric matrices (resp. positive semidefinite matrices). A semidefinite program is a finite-dimensional convex optimization problem $\mathbf{P}$ which in canonical form reads:

$$\mathbf{P}: \quad \rho = \inf_{\mathbf{x}} \left\{ \mathbf{c}^T \mathbf{x} : \sum_{i=1}^{n} \mathbf{A}_j \, x_j \succeq \mathbf{A}_0 \right\} \tag{1.1}$$

where $\mathbf{x} = (x_1, \ldots, x_n) \in \mathbb{R}^n$ and $\mathbf{A}_i \in \mathcal{S}^m$, $i = 0, \ldots, m$. The notation $\mathbf{A} \succeq 0$ stands for the (real symmetric) matrix $\mathbf{A}$ is psd (positive semidefinite), i.e., all its eigenvalues are nonnegative.

The dual $\mathbf{P}^*$ of $\mathbf{P}$ is also a semidefinite program which reads:

$$\mathbf{P}^*: \quad \rho^* = \sup_{\mathbf{X} \in \mathcal{S}^m_+} \left\{ \operatorname{trace}(\mathbf{A}_0 \, \mathbf{X}) : \operatorname{trace}(\mathbf{A}_i \, \mathbf{X}) = c_i, \quad i = 1, \ldots, n \right\}.$$
$$\tag{1.2}$$

Weak duality holds, that is, $\mathbf{c}^T \mathbf{x} \geq \operatorname{trace}(\mathbf{A}_0 \, \mathbf{X})$ for every pair feasible solutions $(\mathbf{x}, \mathbf{X})$ of (1.1) and (1.2).

Theorem 1.1. *Let $\mathbf{A}_i, \mathbf{c}$ be as in* (1.1).

(i) If (1.1) *has a strictly feasible solution then $\rho = \rho^*$. In addition if ρ is finite then* (1.2) *has an optimal solution $\mathbf{X}^* \succeq 0$.*

(ii) If (1.2) *has a positive definite feasible solution $\mathbf{X} \succ 0$ then $\rho = \rho^*$. In addition if ρ is finite then* (1.1) *has an optimal solution $\mathbf{x}^* \in \mathbb{R}^n$.*

(iii) If both (1.1) and (1.2) have a feasible solution $\mathbf{x}$ *and* $\mathbf{X} \succ 0$ *respectively, then* $\rho = \rho^*$ *and both (1.1) and (1.2) have an optimal solution* $\mathbf{x}^*$ *and* $\mathbf{X}^*$ *respectively.*

(iv) If the set of optimal solutions of (1.1) or of (1.2) is compact then $\rho = \rho^*$ *and both are finite.*

Theorem 1.1 is standard while Theorem 1.1(iv) is from Trnovská (2005). The class of semidefinite programs (SDPs) is very important because they have great modeling power and they can be solved efficiently, e.g. by *interior point* algorithms. That is, to the precision $\epsilon > 0$ fixed in advance, (1.1) can be solved in time polynomial in its input size. For more details on SDPs and their applications see e.g. Anjos and Lasserre (2012).

Definition 1.1. A set $\mathcal{X} \subset \mathbb{R}^n$ is said to be semidefinite representable (or is said to have a semidefinite representation) if there exists $m, s \in \mathbb{N}$, $\mathbf{A}_i, \mathbf{B}_j \in \mathcal{S}^m$, $i = 1, \ldots, n$, $j = 0, \ldots, s$, such that:

$$\mathcal{X} = \{ \mathbf{x} : \exists \mathbf{y} \in \mathbb{R}^s \text{ s.t. } \sum_{i=1}^{n} \mathbf{A}_i \, x_i + \sum_{j=1}^{s} \mathbf{B}_j \, y_j \succeq \mathbf{B}_0 \}. \qquad (1.3)$$

In view of what precedes, optimizing a linear function on a semidefinite representable set can be done efficiently by solving a semidefinite program.

1.1.2 *Measures and functions*

Given a closed set $\mathcal{X} \subset \mathbb{R}^\ell$, denote by $\mathcal{B}(\mathcal{X})$ the Borel σ-field of $\mathcal{X}$, $\mathscr{P}(\mathcal{X})$ the space of probability measures on $\mathcal{X}$ and by $\mathscr{B}(\mathcal{X})$ (resp. $\mathscr{C}(\mathcal{X})$) the space of bounded measurable (resp. bounded continuous) functions on $\mathcal{X}$. Let $\mathscr{B}(\mathcal{X})_+$ (resp. $\mathscr{C}(\mathcal{X})_+$) denote the space of nonnegative bounded measurable (resp. continuous) functions on $\mathcal{X}$.

We also denote by $\mathscr{M}(\mathcal{X})$ the space of finite signed Borel measures on $\mathcal{X}$ and by $\mathscr{M}(\mathcal{X})_+$ its subset of finite (positive) measures on $\mathcal{X}$. When $\mathcal{X}$ is compact and $\mathscr{C}(\mathcal{X})$ is equipped with the sup-norm $\mathscr{M}(\mathcal{X})$ is isometrically isomorphic to the topological dual of $\mathscr{C}(\mathcal{X})$ (denoted $\mathscr{M}(\mathcal{X}) \simeq \mathscr{C}(\mathcal{X})^*$).

Given a σ-finite measure μ on $\mathcal{X}$, let $L_1(\mathcal{X}, \mu)$ denote the set of μ-integrable functions on $\mathcal{X}$, that is,

$$L_1(\mathcal{X}, \mu) := \{ f : \int_\mathcal{X} |f| \, d\mu < \infty \}.$$

Definition 1.2. The *support* supp(μ) of a Borel measure μ on $\mathbb{R}^n$ is the (unique) smallest closed set $\mathbf{\Omega} \subset \mathbb{R}^n$ such that $\mu(\mathbb{R}^n \setminus \mathbf{\Omega}) = 0$.

A (signed) measure $\mu \in \mathcal{M}(\mathcal{X})$ with finite support at the points $\mathbf{x}(i) \in \mathbb{R}^n$, $i = 1, \ldots, s$, read $\mu = \sum_{i=1}^{s} \gamma_i \delta_{\mathbf{x}(i)}$ for some real "weights" $\gamma_i \in \mathbb{R}$, $i = 1, \ldots, s$. Such a measure is said to be *atomic* with s atoms.

Absolute continuity. Let μ and λ be two given σ-finite measures on $\mathbb{R}^n$. The notation "$\mu \ll \lambda$" stands for "μ *is absolutely continuous with respect to (w.r.t.)* λ", that is,

$$[B \in \mathcal{B}(\mathbb{R}^n) \quad \text{and} \quad \lambda(B) = 0] \Rightarrow \mu(B) = 0.$$

In such a case, the Radon-Nikodym theorem states that there exists some nonnegative $f \in L_1(\mathbb{R}^n, \lambda)$ such that

$$\mu(B) = \int_B f(\mathbf{x}) \, d\lambda(\mathbf{x}), \quad \forall B \in \mathcal{B}(\mathbb{R}^n),$$

and f is call the Radon-Nikodym derivative of μ w.r.t. λ.

Mutually singular measures. Let μ and λ be two given σ-finite measures on $\mathbb{R}^n$. The notation "$\mu \perp \lambda$" stands for "μ *and* λ *are mutually singular*", that is, there exists $A, B \in \mathcal{B}(\mathbb{R}^n)$ such that $A \cap B = \emptyset$, $A \cup B = \mathbb{R}^n$, and $\mu(B) = \lambda(A) = 0$.

1.1.3 *Polynomials*

Let $\mathbb{R}[\mathbf{x}]$ be the ring of polynomials in the variables $\mathbf{x} = (x_1, \ldots, x_n)$ and let $\mathbb{R}[\mathbf{x}]_d$ be the vector space of polynomials of degree at most d (whose dimension is $s(d) := \binom{n+d}{n}$). Given $\Omega \subset \mathbb{R}^n$, let $\mathcal{P}(\Omega)$ denote the space of polynomials that are nonnegative on Ω. For every $d \in \mathbb{N}$, let $\mathbb{N}_d^n := \{\alpha \in \mathbb{N}^n : |\alpha| \, (= \sum_i \alpha_i) \leq d\}$. A polynomial $p \in \mathbb{R}[\mathbf{x}]_d$ is written

$$\mathbf{x} \mapsto p(\mathbf{x}) = \sum_{\alpha \in \mathbb{N}^n} p_\alpha \mathbf{x}^\alpha,$$

for some vector of coefficients $\mathbf{p} = (p_\alpha) \in \mathbb{R}^{s(d)}$. Denote by $\mathbf{v}_d(\mathbf{x}) = (\mathbf{x}^\alpha) \in \mathbb{R}^{s(d)}$ the vector of monomials up to degree d with the lexicographic ordering; for instance with $n = 3$ and $d = 2$:

$$\mathbf{v}_2(\mathbf{x}) = (1, x_1, x_2, x_3, x_1^2, x_1 x_2, x_1 x_3, x_2^2, x_2 x_3, x_3^2).$$

Denote by $\Sigma[\mathbf{x}]$ (resp. $\Sigma[\mathbf{x}]_d$) the space of sums-of-squares (SOS) polynomials (resp. SOS polynomials of degree at most $2d$), i.e., $p \in \Sigma[\mathbf{x}]_d$ if

$$\mathbf{x} \mapsto p(\mathbf{x}) = \sum_{k=1}^{s} p_k(\mathbf{x})^2,$$

for some $s \in \mathbb{N}$ and some polynomials $p_k \in \mathbb{R}[\mathbf{x}]_d$, $k = 1, \ldots, s$.

Quadratic module. Given a finite family $(g_0, \ldots, g_m) \subset \mathbb{R}[\mathbf{x}]$ (with $g_0(\mathbf{x}) = 1$ for all $\mathbf{x}$) let $Q(g) \subset \mathbb{R}[\mathbf{x}]$ be defined by:

$$Q(g) := \left\{ \sum_{j=0}^{m} \sigma_j \, g_j : \sigma_j \in \Sigma[\mathbf{x}], \, j = 0, \ldots, m \right\} \qquad (1.4)$$

and for $d \in \mathbb{N}$, let $Q_d(g) \subset Q(g)$ be defined by:

$$Q_d(g) := \left\{ \sum_{j=0}^{m} \sigma_j \, g_j : \sigma_j \in \Sigma[\mathbf{x}]; \, \deg \sigma_j g_j \leq 2d, \, j = 0, \ldots, m \right\}. \qquad (1.5)$$

> Importantly and crucial for the Moment-SOS hierarchy is the fact that testing membership in $Q_d(g)$ can be done by solving a semidefinite program. Equivalently, $Q_d(g)$ has a semidefinite representation (see Definition 1.1).

Definition 1.3. The set $Q(g)$ in (1.4) is called the *quadratic module* generated by $(g_1, \ldots, g_m) \subset \mathbb{R}[\mathbf{x}]$. It is said to be Archimedean if there exists $M > 0$ such that the quadratic polynomial $\mathbf{x} \mapsto M - \|\mathbf{x}\|^2$ belongs to $Q(g)$. The set $Q_d(g)$ in (1.5) is a *truncated* version of $Q(g)$.

If $Q(g)$ is Archimedean then $\mathbf{K}$ is compact because $\mathbf{x} \mapsto M - \|\mathbf{x}\|^2 \in Q(g)$ implies $\|\mathbf{x}\|^2 \leq M$ for all $\mathbf{x} \in \mathbf{K}$. Therefore the fact that $\mathbf{x} \mapsto M - \|\mathbf{x}\|^2 \in Q(g)$ can be interpreted as an *algebraic certificate* that $\mathbf{K}$ is compact. On the other hand, in general, $\mathbf{K}$ compact does not imply that $Q(g)$ is Archimedean.

Lemma 1.2. *Let $g_j \in \mathbb{R}[\mathbf{x}]$, $j = 1, \ldots, m$, and let $\mathbf{X} := \{ \mathbf{x} : g_j(\mathbf{x}) \geq 0, j = 1, \ldots, m \}$ be with nonempty interior and such that $Q(g)$ is Archimedean. Then $Q_d(g)$ is closed.*

Moment matrix. Given a sequence $\mathbf{y} = (y_\alpha)_{\alpha \in \mathbb{N}^n}$, let $L_\mathbf{y} : \mathbb{R}[\mathbf{x}] \to \mathbb{R}$ be the linear (Riesz) functional

$$f \left(= \sum_\alpha f_\alpha \mathbf{x}^\alpha\right) \mapsto L_\mathbf{y}(f) := \sum_\alpha f_\alpha y_\alpha. \tag{1.6}$$

Given $\mathbf{y}$ and $d \in \mathbb{N}$, the *moment* matrix associated with $\mathbf{y}$, is the real symmetric $s(d) \times s(d)$ matrix $\mathbf{M}_d(\mathbf{y})$ with rows and columns indexed in $\mathbb{N}^n_d$ and with entries

$$\mathbf{M}_d(\mathbf{y})(\alpha, \beta) := L_\mathbf{y}(\mathbf{x}^{\alpha+\beta}) = y_{\alpha+\beta}, \quad \alpha, \beta \in \mathbb{N}^n_d.$$

Definition 1.4. A sequence $\mathbf{y} = (y_\alpha)_{\alpha \in \mathbb{N}^n}$ is said to have a *representing measure* μ if there exists a Borel measure μ on $\mathbb{R}^n$ such that $y_\alpha = \int \mathbf{x}^\alpha \, d\mu$ for all $\alpha \in \mathbb{N}^n$. If μ is unique then μ said to be *moment determinate*. (Every measure with compact support is moment-determinate.)

A necessary condition for the existence of such a μ is that $\mathbf{M}_d(\mathbf{y}) \succeq 0$ for all d. Equivalently $L_\mathbf{y}(f^2) \geq 0$ for all $f \in \mathbb{R}[\mathbf{x}]$. But it is only a sufficient condition (except in the univariate case $n = 1$).

In the rest of the book we will use repeatedly the following useful property of the moment matrix.

Lemma 1.3. *Let* $\mathbf{y} \in \mathbb{R}^{s(2d)}$ *be such that* $\mathbf{M}_d(\mathbf{y}) \succeq 0$. *Then*

$$\sup_{\alpha \in \mathbb{N}^n_{2d}} |y_\alpha| \leq \max[\, L_\mathbf{y}(1); \max_{i=1,\dots,n}[\, L_\mathbf{y}(x_i^{2d})\,]\,]. \tag{1.7}$$

Localizing matrix. Given a sequence $\mathbf{y} = (y_\alpha)_{\alpha \in \mathbb{N}^n}$, and a polynomial $g \in \mathbb{R}[\mathbf{x}]$, the *localizing* moment matrix associated with $\mathbf{y}$ and g, is the real symmetric $s(d) \times s(d)$ matrix $\mathbf{M}_d(g \, \mathbf{y})$ with rows and columns indexed in $\mathbb{N}^n_d$ and with entries

$$\mathbf{M}_d(g \, \mathbf{y})(\alpha, \beta) := L_\mathbf{y}(g(\mathbf{x}) \, \mathbf{x}^{\alpha+\beta})$$
$$= \sum_\gamma g_\gamma \, y_{\alpha+\beta+\gamma}, \quad \alpha, \beta \in \mathbb{N}^n_d.$$

1.2 Some background results

Disintegration of a measure. Given a probability measure μ on a cartesian product $\mathcal{X} \times \mathcal{Y}$ of topological spaces, we may decompose μ into its

marginal $\mu_{\mathbf{x}}$ on $\mathcal{X}$ and a stochastic kernel (or conditional probability measure) $\hat{\mu}(d\mathbf{y}|\cdot)$ on $\mathcal{Y}$ given $\mathcal{X}$, and where:
- For every $\mathbf{x} \in \mathcal{X}$, $\hat{\mu}(d\mathbf{y}|\mathbf{x}) \in \mathscr{P}(\mathcal{Y})$, and
- For every $B \in \mathcal{B}(\mathcal{Y})$, the function $\mathbf{x} \mapsto \hat{\mu}(B|\mathbf{x})$ is measurable.

Then

$$\mu(A \times B) = \int_A \hat{\mu}(B|\mathbf{x})\,\mu_{\mathbf{x}}(dx), \quad \forall A \in \mathcal{B}(\mathcal{X}), B \in \mathcal{B}(\mathcal{Y}).$$

The K-moment and Generalized Moment Problems

Definition 1.5. (i) The **K-Moment Problem**: Given a closed set $\mathbf{K} \subset \mathbb{R}^n$ and a real sequence $\mathbf{y} = (y_\alpha)_{\alpha \in \mathbb{N}^n}$, the **K**-moment problem is to decide whether there exists $\mu \in \mathscr{M}_+(\mathbf{K})$ such that

$$y_\alpha = \int_{\mathbf{K}} \mathbf{x}^\alpha\,d\mu(\mathbf{x}), \quad \forall \alpha \in \mathbb{N}^n. \tag{1.8}$$

The *truncated* version of the **K**-moment problem asks for the same question except now $\mathbb{N}^n$ in (1.8) is replaced by a (finite) subset $\Gamma \subset \mathbb{N}^n$.

(ii) Let $\Theta \subset \mathbb{N}$, I be a finite set, $(n_i)_{i \in I} \subset \mathbb{N}$ and $\mathbf{K}_i \subset \mathbb{R}^{n_i}$, $i \in I$. Let $f_{k,i} : \mathbb{R}^{n_i} \to \mathbb{R}$ be given and $\mathbf{b} = (b_k)_{k \in \Theta}$ a real vector. The **Generalized Moment Problem (GMP)** is the infinite-dimensional LP

$$\inf_{\mu_i \in \mathscr{M}(\mathbf{K}_i)_+} \Big\{ \sum_{i \in I} \int_{\mathbf{K}_i} f_{0,i}\,d\mu_i : \sum_{i \in I} \int_{\mathbf{K}_i} f_{k,i}\,d\mu_i \;\;\Delta\;\; b_k, \quad k \in \Theta \Big\} \tag{1.9}$$

where $\Delta \in \{``="", ``\geq""\}$.

1.2.1 *Conditions for existence of a representing measure*

For existence of a representing measure, the following sufficient conditions in Lasserre (2015b)[Proposition 2.37] are very useful:

Lemma 1.4. *Let* $\mathbf{y} = (y_\alpha)_{\alpha \in \mathbb{N}^n}$ *satisfy* $\mathbf{M}_d(\mathbf{y}) \succeq 0$ *for all* $d = 0, 1, \ldots$
    (i) If there exists $M, a > 0$ *such that:*

$$|y_\alpha| < M\,a^{|\alpha|}, \quad \forall \alpha \in \mathbb{N}^n, \tag{1.10}$$

then $\mathbf{y}$ *has a representing measure supported on* $[-a, a]^n$, *and in addition μ is moment determinate.*

   (ii) If

$$\sum_{k=1}^{\infty} L_{\mathbf{y}}(x_i^{2k})^{-1/2k} = +\infty, \qquad i = 1, \ldots, n, \qquad (1.11)$$

then $\mathbf{y}$ *has a representing measure on* $\mathbb{R}^n$, *and in addition* μ *is moment determinate.*

Condition (1.11) due to Nussbaum is the multivariate generalization of its earlier univariate version due to Carleman; see e.g. Lasserre (2010).

We next describe some sufficient conditions for existence of a representing measure, based on finitely many moments only. The following result of Fialkow (2016) is a consequence of Blekherman (2015).

Theorem 1.5. *Let* $\mathbf{y} = (y_\alpha)_{\alpha \in \mathbb{N}_{2d}^n}$ *be such that* $\mathbf{M}_d(\mathbf{y}) \succeq 0$. *If* $\operatorname{rank} \mathbf{M}_d(\mathbf{y}) \leq 3d - 3$ *(if* $d > 2$) *or* $\operatorname{rank} \mathbf{M}_d(\mathbf{y}) \leq 6$ *(if* $d = 2$) *then the sequence* $\mathbf{y} = (y_\alpha)_{\alpha \in \mathbb{N}_{2d-1}^n}$ *has an atomic representing measure with at most* $\operatorname{rank} \mathbf{M}_d(\mathbf{y})$ *atoms.*

Another very useful condition is the following result of Curto and Fialkow (see e.g. Lasserre (2010)):

Theorem 1.6. *Let* $g_j \in \mathbb{R}[\mathbf{x}]$ *and* $d_j := \lceil (\deg g_j)/2 \rceil$, $j = 1, \ldots, m$. *Let* $\mathbf{y} = (y_\alpha)_{\alpha \in \mathbb{N}_{2d}^n}$ *be such that* $\mathbf{M}_d(\mathbf{y}) \succeq 0$ *and* $\mathbf{M}_{d-d_j}(g_j \mathbf{y}) \succeq 0$ *for all* $j = 1, \ldots, m$. *If*

$$\operatorname{rank} \mathbf{M}_d(\mathbf{y}) = \operatorname{rank} \mathbf{M}_{d-d^*}(\mathbf{y}) \ (=: s) \qquad (1.12)$$

(where $d^* = \max_j d_j$) *then* $\mathbf{y}$ *has a representing (atomic) measure* μ *whose support consists of* s *atoms contained in the set* $\{\mathbf{x} : g_j(\mathbf{x}) \geq 0, j = 1, \ldots, m\}$. *In addition, the atoms and weights of* μ *can be "extracted" from the sequence* $\mathbf{y}$ *by a linear algebra routine.*

For details on an algorithm to extract the atoms and weights of μ in Theorem 1.6, the reader is referred to Henrion and Lasserre (2005) and Henrion et al. (2009).

Lemma 1.7. *Let* $\mathbf{K} \subset \mathbb{R}^n$ *be a compact set and with* $s(d) := \binom{n+d}{n}$, *let:*

$$\mathcal{M}(\mathbf{K})_d := \{ \mathbf{y} = (y_\alpha) \in \mathbb{R}^{s(d)} : \mathbf{y} \text{ has a representing measure on } \mathbf{K} \}. \qquad (1.13)$$

Then $\mathcal{M}(\mathbf{K})_d$ *is a convex cone whose dual cone is:*

$$\mathcal{M}(\mathbf{K})_d^* = \mathcal{P}(\mathbf{K})_d := \{\, p \in \mathbb{R}[\mathbf{x}]_d : p \text{ is nonnegative on } \mathbf{K}\,\}. \quad (1.14)$$

In addition, $\mathcal{P}(\mathbf{K})_d^* = \mathcal{M}(\mathbf{K})_d$.

For a proof see e.g. Lasserre (2015a). Next let $g_j \in \mathbb{R}[\mathbf{x}]$, $j = 0, \ldots, m$, with $g_0(\mathbf{x}) = 1$ and $g_1(\mathbf{x}) = M - \|\mathbf{x}\|^2$ for all $\mathbf{x}$ and some $M > 0$. Let $d_j := \lceil \deg(g_j)/2 \rceil$, $j = 0, \ldots, m$, and let:

$$\mathbf{K} := \{\, \mathbf{x} \in \mathbb{R}^n : g_j(\mathbf{x}) \geq 0, \; j = 1, \ldots, m\,\}. \quad (1.15)$$

Such a set $\mathbf{K}$ is called a closed *basic semi-algebraic set.* Next, with $d \geq \underline{d} := \max_j d_j$, let

$$\mathcal{M}(\mathbf{K})_{2d}^{SDP} := \{\, \mathbf{y} \in \mathbb{R}^{s(2d)} : y_0 = 1; \\ \mathbf{M}_{d-d_j}(g_j\,\mathbf{y}) \succeq 0, \; j = 0, \ldots, m\,\}. \quad (1.16)$$

Lemma 1.8. *Let* $\mathbf{K}$ *be as in (1.15) with* $g_1(\mathbf{x}) = M - \|\mathbf{x}\|^2$. *Then the set* $\mathcal{M}(\mathbf{K})_{2d}^{SDP}$ *defined in (1.16) is compact and*

$$\mathcal{M}(\mathbf{K})_{2d} \subset \mathcal{M}(\mathbf{K})_{2d}^{SDP}, \quad \forall d \geq \underline{d}.$$

Proof. It suffices to prove that $\mathcal{M}(\mathbf{K})_{2d}^{SDP}$ is bounded. Using $\mathbf{M}_d(g_1\,\mathbf{y}) \succeq 0$ and the definition of g_1 yields $L_{\mathbf{y}}(x_i^{2k}(M - \|\mathbf{x}\|^2)) \geq 0$ for all $0 \leq k \leq d - 1$, and all $i = 1, \ldots, n$. In particular this yields $M \geq L_{\mathbf{y}}(x_i^2)$ for all $i = 1, \ldots, n$. Iterating yields $M^2 \geq M L_{\mathbf{y}}(x_i^2) \geq L_{\mathbf{y}}(x_i^4)$ for all i, to finally obtain $L_{\mathbf{y}}(x_i^{2d}) \leq M^d$ for all $i = 1, \ldots, n$. Combining with Lemma 1.3 yields the first desired result. The second statement is straightforward. $\square$

1.2.2 *Putinar's Certificate of Positivity*

Putinar's Theorem in Putinar (1993) solves the $\mathbf{K}$-moment problem for a class of compact basic semi-algebraic sets. It provides a certificate of positivity on $\mathbf{K} \subset \mathbb{R}^n$ which is crucial to the convergence of the numerical scheme underlying the Moment-SOS Hierarchy.

With $g_1, \ldots, g_m \in \mathbb{R}[\mathbf{x}]$ and $g_0(\mathbf{x}) = 1$ for all $\mathbf{x}$, let $\mathbf{K}$ be as in (1.15).

Theorem 1.9 (Putinar's Positivstellensatz). *Let* $\mathbf{K} \subset \mathbb{R}^n$ *as in (1.15) be nonempty and assume that* $Q(g)$ *in (1.4) is Archimedean (hence* $\mathbf{K}$ *is compact).*

(i) *If* $f \in \mathbb{R}[\mathbf{x}]$ *is (strictly) positive on* $\mathbf{K}$ *then* $f \in Q(g)$, *i.e.,*

$$f = \sum_{j=0}^{m} \sigma_i \, g_j, \qquad (1.17)$$

for some SOS polynomials σ_j, $j = 0, \ldots, m$.
(ii) *If* $\mathbf{y} = (y_\alpha)_{\alpha \in \mathbb{N}^n}$ *is a real sequence such that*

$$\mathbf{M}_d(g_j \, \mathbf{y}) \succeq 0, \qquad 0 \le j \le m; \ d = 0, 1, \ldots \qquad (1.18)$$

then $\mathbf{y}$ *has a representing measure* μ *supported on* $\mathbf{K}$.

Theorem 1.9(i) is the algebraic facet of Putinar's Theorem and is concerned with the representation of polynomials positive on $\mathbf{K}$. Theorem 1.9(ii) is its functional analysis facet and is concerned with the $\mathbf{K}$-moment problem. In other words, Theorem 1.9(i)-(ii) is a beautiful and elegant illustration of the duality between moments and positive polynomials. It is in fact a refinement and simplification of Schmüdgen's theorem in Schmüdgen (1991) in which all products of the g_j's are involved (but with no Archimedean assumption required).

1.2.3 Stokes' Theorem

Stokes' Theorem can be viewed as the multivariate generalization of standard integration by parts. It relates an integral on a smooth manifold Ω with one on its boundary $\partial \Omega$.

Theorem 1.10. *Let* $\Omega \subset \mathbb{R}^n$ *be a closed set with smooth boundary* $\partial \Omega$, *let* X *be a vector field and* $f \in C^1(\Omega)$. *Then:*

$$\int_\Omega \mathrm{Div}(X) \, f(\mathbf{x}) \, d\mathbf{x} + \int_\Omega \langle X, \nabla f(\mathbf{x}) \rangle \, d\mathbf{x} = \int_{\partial \Omega} \langle \vec{n}_\mathbf{x}, X \rangle \, f(\mathbf{x}) \, d\sigma(\mathbf{x}), \quad (1.19)$$

where Div *is the Divergence operator,* $\vec{n}_\mathbf{x}$ *is the outward pointing normal at the point* $\mathbf{x} \in \partial \Omega$, *and* σ *is the* $n - 1$*-dimensional Hausdorff measure on* $\partial \Omega$. *For instance, let* $\mathbf{X} := e_i$ *where* $e_i \in \mathbb{R}^n$ *with* $e_{ik} = \delta_{i=k}$, $k = 1, \ldots, n$. *Then* (1.19) *reads:*

$$\int_\Omega \frac{\partial f(\mathbf{x})}{\partial x_i} \, d\mathbf{x} = \int_{\partial \Omega} \langle \vec{n}_\mathbf{x}, e_i \rangle \, f(\mathbf{x}) \, d\sigma(\mathbf{x}), \qquad (1.20)$$

See e.g. Taylor (1996)[Proposition 2.3, p. 128]. Stokes' Theorem has been generalized to rough domains Ω (e.g. with corners) in Whitney (1957)[Theorem 14A].

1.2.4   *The Christoffel function*

Given a measure μ whose support is a compact set $\Omega \subset \mathbb{R}^n$ and such that its moment matrix $\mathbf{M}_d(\mu)$ is positive definite for all d, the *Christoffel* function $c_d : \mathbb{R}^n \to \mathbb{R}_+$ is defined by:

$$\mathbf{x} \mapsto c_d(\mathbf{x}) := \frac{1}{\mathbf{v}_d(\mathbf{x})^T \mathbf{M}_d(\mu)^{-1} \mathbf{v}_d(\mathbf{x})}, \qquad \mathbf{x} \in \mathbb{R}^n,$$

where $\mathbf{v}_d(\mathbf{x}) = (\mathbf{x}^\alpha)_{\alpha \in \mathbb{N}_d^n} \in \mathbb{R}^{s(d)}$. So we may and will call the SOS polynomial $\mathbf{x} \mapsto c_d(\mathbf{x})^{-1} = \mathbf{v}_d(\mathbf{x})^T \mathbf{M}_d(\mu)^{-1} \mathbf{v}_d(\mathbf{x})$ the *Christoffel* polynomial associated with μ. It turns out that $c_d(\mathbf{x})^{-1} = \sum_{|\alpha| \leq d} P_\alpha(\mathbf{x})^2$, where $(P_\alpha)_{\alpha \in \mathbb{N}^n}$ is a family of orthonormal polynomials w.r.t. μ.

The Christoffel function also has the extremal characterization:

$$c_d(\boldsymbol{\xi}) = \min_{p \in \mathbb{R}[\mathbf{x}]_d} \left\{ \int_\Omega p(\mathbf{x})^2 \, d\mu(\mathbf{x}) : p(\boldsymbol{\xi}) = 1 \right\}, \quad \boldsymbol{\xi} \in \mathbb{R}^n,$$

and is well known in theory of approximation. In particular, its asymptotic properties as $d \to \infty$ has been a topic of investigation for a long time with exact characterization in a few cases only. However,

$$\forall \mathbf{x} \notin \Omega : \quad \lim_{d \to \infty} \binom{n+d}{n} c_d(\mathbf{x}) = 0$$

whereas

$$\int_\Omega \left(\binom{n+d}{n} c_d \right)^{-1} d\mu = 1,$$

which shows that somehow $c_d(\mathbf{x})$ "identifies" the support of μ.

1.3   Notes and sources

For more details on basic background results related to measures, moments, positive polynomials, algebraic geometry, convex algebraic geometry, and "polynomial optimization", the interested reader is referred to Anjos and Lasserre (2012), Basu et al. (2003), Blekherman et al. (2013),

Fialkow (2016), Lasserre (2010), Lasserre (2015b), Prestel (2001), Putinar and Sullivant (2009), and Schmüdgen (2017).

Bibliography

Anjos, M., and Lasserre, J. B. (2012). *Handbook on Semidefinite, Conic an Polynomial Optimization*, (Eds.) (Springer, New York).

Basu, S., Pollack, R., Roy, M.-F. (2003). *Algorithms in Real Algebraic Geometry* (Springer, Berlin).

Blekherman, G., Parrilo, P. A., Thomas, R. (2013). *Semidefinite Optimization and Convex Algebraic Geometry* (MOS-SIAM Series, SIAM, Philadelphia).

Blekherman G. (2015). *Positive Gorenstein ideals*, Proc. Amer. Math. Soc. **143**, pp. 69–86.

Fialkow L. A. (2016). The truncated K-moment problem: A survey, in: *Operator Theory: the State of the Art*, Dumitru Gaspar, Marius Junge, Dan Timotin, Florian-Horia Vasilescu (Eds.), Conference Proceedings, Timisoara, June 30–July 5, 2014, Theta Foundation, pp. 25–51.

Henrion, D., and Lasserre J. B. (2005). *Detecting Global Optimality and Extracting Solutions in GloptiPoly*, in *Positive Polynomials in Control*, D. Henrion and A. Garulli (Eds.), Lecture Notes in Control and Information Sciences, Springer-Verlag, Berlin 2005), pp. 293–300.

Henrion D., Lasserre, J. B., Lofberg J. (2009). *Gloptipoly 3: moments, optimization and semidefinite programming*, Optim. Methods and Softwares **24**, pp. 761–779.

Lasserre, J. B. (2010). *Moments, Positive Polynomials and Their Applications* (Imperial College Press, London).

Lasserre, J. B. (2015a). *A generalization of Löwner-John's ellipsoid theorem*, Math. Program. **152**, pp. 559–591.

Lasserre, J. B. (2015b). *Introduction to Polynomial and Semi-Algebraic Optimization* (Cambridge University Press, Cambridge, UK).

Prestel, A., and Delzell, C. N. (2001). *Positive Polynomials* (Springer, Berlin).

Putinar, M. (1993). *Positive polynomials on compact semi-algebraic sets*, Indiana Univ. Math. J. **42**, pp. 969–984.

Putinar, M., and Sullivant, S. (2009). *Emerging Applications of Algebraic Geometry*, (Eds.) (Springer-Verlag, New York).

Schmüdgen, K. (1991). *The K-moment problem for compact semi-algebraic sets*, Math. Ann. **289**, 203–206.

Schmüdgen, K. (2017). *The Moment Problem* (Springer, New York).

Taylor, M. E. (1996). *Partial Differential Equations: Basic Theory* (Springer Texts in Mathematics, Springer-Verlag, New York, Inc.).

Trnovská, M. (2005). *Strong duality conditions in semidefinite programming*, J. Elec. Eng. **56**, pp. 1–5.

Whitney, H. (1957). *Geometric Integration Theory* (Princeton University Press, Princeton).

Chapter 2

Principle of the Moment-SOS Hierarchy

We describe the underlying principle behind the Moment-SOS hierarchy and provide a brief illustration in polynomial optimization.

2.1 The Moment-SOS hierarchy

The diagram in Figure 2.1 below is a brief summary of the Moment-SOS hierarchy. It consists of four steps described in informal terms. Then for illustration purpose, in the next section we briefly describe those four steps in the context of solving (global) polynomial optimization problems (which was the initial and main motivation when the Moment-SOS hierarchy was introduced in Lasserre (2000)).

Step 1

Start from an initial problem **P** which is non-convex and described with algebraic and semi-algebraic data.

Step 2

Provide an associated *infinite-dimensional linear program* on an appropriate space of *measures* with supports on given compact semi-algebraic sets. Under some conditions, this (primal) LP is equivalent to the initial problem **P**, that is, it has same optimal value and from an optimal solution of the LP one may recover an optimal solution of **P**. By standard LP-duality, a dual LP is formulated in appropriate spaces of continuous functions. The constraints of this LP are positivity of some continuous functions on some semi-algebraic sets.

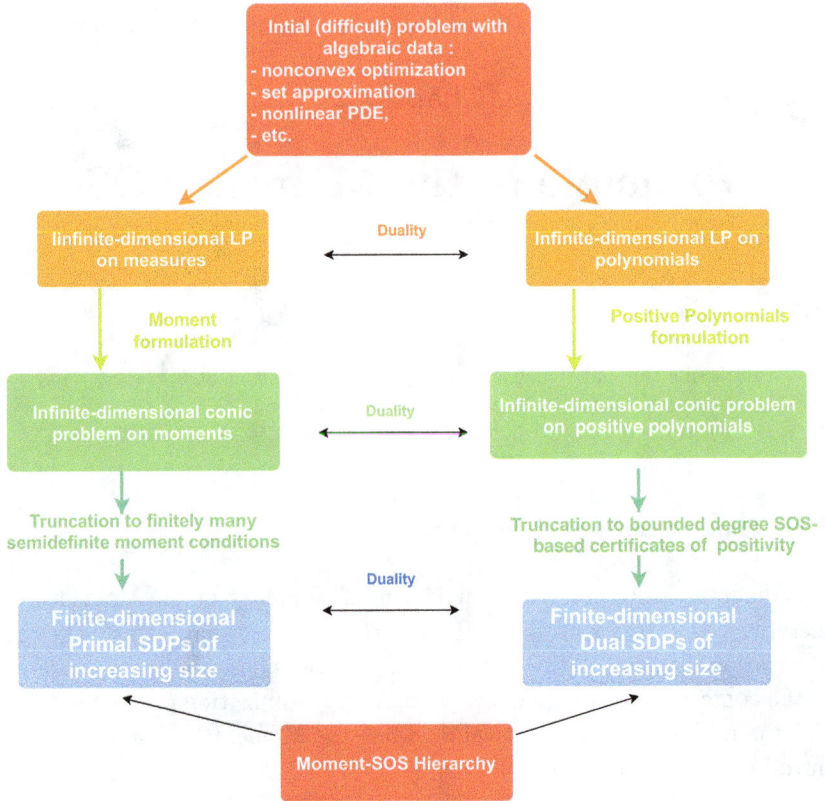

Fig. 2.1   The Moment-SOS Hierarchy

Step 3

Using the algebraic features of data of the initial problem $\mathbf{P}$, the primal LP is equivalent to a convex conic program on *moments* of the unknown measures. The unknown are infinite sequences of reals and conic constraints state that such infinite sequences are indeed moments of (unknown) measures on the required semi-algebraic supports.

Similarly, the dual LP is a convex conic optimization problem on unknown polynomials. The conic constraints state that such polynomials must be positive on the previous semi-algebraic sets.

Step 4

By truncation to finitely many moments, the infinite-dimensional conic optimization problem is replaced by a hierarchy of finite-dimensional conic optimization problems of increasing size, and its associated hierarchy of dual problems. Convergence of the hierarchy is guaranteed by invoking powerful results on positivity certificates for polynomials on compact sets.

Depending on which positivity certificate is used one obtains different types of conic problems. For instance if one uses the Sum-of-Squares (SOS) based Putinar's PositistivStellensatz (Theorem 1.9), each conic problem in the hierarchy is a semidefinite program and one thus obtains the Moment-SOS hierarchy. On the other hand if one uses Krivine-Stengle positivity certificates, one obtains a hierarchy of LPs.

2.2 Illustration in polynomial optimization

In Step 1, one considers the non-convex global optimization problem:

$$\mathbf{P}: \quad f^* = \min_{\mathbf{x}} \{ f(\mathbf{x}) : \mathbf{x} \in \mathbf{K} \} \tag{2.1}$$

where $\mathbf{K} := \{ \mathbf{x} \in \mathbb{R}^n : g_j(\mathbf{x}) \geq 0, \ j = 1, \ldots, m \}$ and $f, g_j \in \mathbb{R}[\mathbf{x}]$, $j = 1, \ldots, m$. Notice that all data of $\mathbf{P}$ are algebraic since the criterion f is a polynomial and the feasible set $\mathbf{K} \subset \mathbb{R}^n$ is semi-algebraic. We assume that $\mathbf{K}$ is compact and the polynomials g_j's satisfy the hypotheses of Theorem 1.9.

Step 2

$\mathbf{P}$ in (2.1) is equivalent to:

$$f^* = \inf_{\mu \in \mathscr{M}(\mathbf{K})_+} \{ \int_{\mathbf{K}} f \, d\mu : \mu(\mathbf{K}) = 1 \}, \tag{2.2}$$

which is the infinite-dimensional LP on measures alluded to in Step 2 of Figure 2.1. It is a conic optimization problem on the space $\mathscr{M}(\mathbf{K})$ of finite signed Borel measures on $\mathbf{K}$ and with convex cone $\mathscr{M}(\mathbf{K})_+$ of positive Borel measures on $\mathbf{K}$. But observe that we also have:

$$f^* = \sup_{f \in \mathscr{C}(\mathbf{K})} \{\lambda : f - \lambda \in \mathscr{C}(\mathbf{K})_+\}, \tag{2.3}$$

which is the infinite-dimensional LP on functions alluded to in Step 2 and is a conic optimization problem on $\mathscr{C}(\mathbf{K})$ (the space of continuous functions on $\mathbf{K}$) and the convex cone $\mathscr{C}(\mathbf{K})_+$ of nonnegative continuous functions on $\mathbf{K}$. The conic constraint $f - \lambda \in \mathscr{C}(\mathbf{K})_+$ is a positivity constraint on the compact basic semi-algebraic set $\mathbf{K}$.

Importantly, the rephrasing of $\mathbf{P}$ in the form (2.2) or (2.3) is valid only for the global minimum f^* (it is *not* valid for a local minimum) and by standard duality in Linear Programming, both problems (2.2) and (2.3) are dual of each other.

Also note that so far we have nowhere used the fact that $\mathbf{P}$ has algebraic data. Indeed both (2.2) and (2.3) are completely equivalent to $\mathbf{P}$ even if f and the g_j's would *not* be polynomials. On the other hand, this algebraic feature of $\mathbf{P}$ is crucial in the next Step 3 and Step 4. Otherwise both (2.2) and (2.3) remain only a (sterile) rephrasing of $\mathbf{P}$.

Step 3

Using the fact that f is a polynomial of degree, say d (i.e., $f \in \mathbb{R}[\mathbf{x}]_d$), and writing $f = \sum_{\alpha \in \mathbb{N}^n_d} f_\alpha \mathbf{x}^\alpha$ in the usual monomial basis $(\mathbf{x}^\alpha)_{\alpha \in \mathbb{N}^n}$,

$$\int_{\mathbf{K}} f d\mu = \sum_{\alpha \in \mathbb{N}^n_d} f_\alpha \underbrace{\int_{\mathbf{K}} \mathbf{x}^\alpha d\mu}_{= y_\alpha} = \sum_{\alpha \in \mathbb{N}^n_d} f_\alpha y_\alpha = L_{\mathbf{y}}(f),$$

where $L_{\mathbf{y}}$ is the Riesz functional introduced in (1.6). Hence the conic problem (2.2) also reads:

$$f^* = \inf_{\mathbf{y} \in \Theta} \{ L_{\mathbf{y}}(f) : L_{\mathbf{y}}(1) = 1 \}, \tag{2.4}$$

where Θ is the convex cone defined by:

$$\Theta = \{ \mathbf{y} \in \mathbb{N}^n : \exists \mu \in \mathscr{M}(\mathbf{K})_+ \text{ s.t. } y_\alpha = \int_{\mathbf{K}} \mathbf{x}^\alpha d\mu, \quad \alpha \in \mathbb{N}^n \}.$$

The dual of (2.4) reads

$$\sup_\lambda \{ \lambda : f - \lambda \in \mathscr{P}(\mathbf{K})_+ \}, \tag{2.5}$$

where $\mathscr{P}(\mathbf{K})_+$ is the convex cone of polynomials positive on $\mathbf{K}$ (and $\mathscr{M}(\mathbf{K})_+ = \mathscr{P}(\mathbf{K})_+^*$ whenever $\mathbf{K}$ is compact).

Step 4

Let $d_j = \lceil \deg(g_j)/2 \rceil$, $j = 1, \ldots, m$ and $d_f = \lceil \deg(f)/2 \rceil$. By restricting to finite "moment" sequences $\mathbf{y} = (y_\alpha)_{\alpha \in \mathbb{N}_{2d}^n}$ one replaces the infinite-dimensional LP (2.4) with the hierarchy of finite-dimensional semidefinite programs:

$$\mathbf{P}_d: \quad \rho_d = \inf_{\mathbf{y} \in \mathbb{N}_{2d}^n} \{ L_{\mathbf{y}}(f) : L_{\mathbf{y}}(1) = 1 \\ \mathbf{M}_d(\mathbf{y}), \mathbf{M}_{d-d_j}(g_j\,\mathbf{y}) \succeq 0, j = 1, \ldots, m \}, \quad (2.6)$$

indexed by $\max[d_f, \max_j d_j] \leq d \in \mathbb{N}$. For each fixed $d \in \mathbb{N}$ the semidefiniteness conditions $\mathbf{M}_d(\mathbf{y})$, $\mathbf{M}_{d-d_j}(g_j\,\mathbf{y}) \succeq 0$, $j = 1, \ldots, m$, are necessary conditions for the finite sequence $\mathbf{y} = (y_\alpha)_{\alpha \in \mathbb{N}_{2d}^n}$ to be moments up to order $2d$ of a measure supported on $\mathbf{K}$. So (2.6) is a *relaxation* of (2.2) and therefore $\rho_d \leq f^*$ for all $d \in \mathbb{N}$.

Letting $g_0(\mathbf{x}) := 1$ for all $\mathbf{x}$ and $d_0 = 0$, the dual of (2.6) reads:

$$\mathbf{P}_d^*: \quad \rho_d^* = \sup_{\sigma_j, \lambda} \{ \lambda : f - \lambda = \sum_{j=0}^m \sigma_j\,g_j \\ \sigma_j \in \Sigma[\mathbf{x}]_{d-d_j}, j = 0, \ldots, m \}. \quad (2.7)$$

By weak duality $\rho_d^* \leq \rho_d$ for all d and (2.7) is a *reinforcement* of (2.3). Indeed one has replaced the difficult positivity constraint $f - \lambda \in \mathscr{P}(\mathbf{K})_+$ in (2.5) with the more tractable condition

$$f - \lambda = \sum_{j=0}^m \sigma_j\,g_j; \quad \sigma_j \in \Sigma[\mathbf{x}]_{d-d_j}, \quad (2.8)$$

which states that $f - \lambda$ is positive on $\mathbf{K}$ with a SOS-based positivity certificate derived from Putinar's representation of polynomials that are strictly positive on $\mathbf{K}$; see Theorem 1.9. Constraint (2.8) is tractable because it translates into a semidefinite constraint on λ and the coefficients of the SOS polynomials (σ_j). If $\operatorname{int}(\mathbf{K}) \neq \emptyset$ or if the set of optimal solutions of (2.6) is compact (then invoking Trnovská (2005)), there is no duality gap between (2.6) and (2.7), i.e., $\rho_d = \rho_d^*$ for all d.

Facts

• By invoking Theorem (1.9)(ii) it follows that $\rho_d \uparrow f^*$ as $d \to \infty$.

• By a result of Nie (2014), it turns out that generically the convergence $\rho_d \uparrow f^*$ is finite, that is, $f^* = \rho_d$ for some finite d^*. When the global minimizer $\mathbf{x}^* \in \mathbf{K}$ is unique then an optimal solution $\mathbf{y}^*$ of the semidefinite program $\mathbf{P}_{d^*}$ is the sequence of moments up to order $2d^*$ of the Dirac measure $\mu^* = \delta_{\mathbf{x}^*}$; therefore, $L_{\mathbf{y}^*}(x_i) = x_i^*$ for all $i = 1, \ldots, n$. That is, one *reads* the global minimizer $\mathbf{x}^* = (x_1^*, \ldots, x_n^*) \in \mathbf{K}$ from the n first-moments in the finite sequence $\mathbf{y}^*$.

• When there are finitely many global minimizers $\mathbf{x}^* \in \mathbf{K}$ and if a sufficient (flat) rank-condition is satisfied for $\mathbf{M}_d(\mathbf{y}^*)$ at an optimal solution $\mathbf{y}^*$ of $\mathbf{P}_d$, then one may extract $\text{rank}\,\mathbf{M}_d(\mathbf{y}^*)$ global minimizers by a linear algebra routine; see Henrion and Lasserre (2005) and Nie (2013).

2.3 Notes and sources

Most of this chapter is from Lasserre (2010, 2015b) in which the reader will find a lot of additional material with appropriate references and a description of several applications of the Moment-SOS hierarchy; see also the more recent tutorial in Lasserre (2018). In particular, the Moment-SOS Hierarchy is implemented in the GloptiPoly software of Henrion et al. (2009) available at: http://homepages.laas.fr/henrion/software/gloptipoly3/

Bibliography

Henrion, D., and Lasserre J. B. (2005). *Detecting Global Optimality and Extracting Solutions in GloptiPoly*, in *Positive Polynomials in Control*, D. Henrion and A. Garulli (Eds.), Lecture Notes in Control and Information Sciences, Springer-Verlag, Berlin 2005), pp. 293–300.

Henrion D., Lasserre, J. B., Lofberg J. (2009). *Gloptipoly 3: moments, optimization and semidefinite programming*, Optim. Methods and Softwares **24**, pp. 761–779.

Lasserre, J. B. (2010). *Moments, Positive Polynomials and Their Applications* (Imperial College Press, London).

Lasserre, J. B. (2015b). *Introduction to Polynomial and Semi-Algebraic Optimization* (Cambridge University Press, Cambridge, UK).

Lasserre, J. B. (2000). *Optimisation globale et théorie des moments*, C. R. Acad. Sci. Paris **331**, Série 1, pp. 929–934.

Lasserre, J. B. (2018). *The Moment-SOS hierarchy*, Proceedings of the International Congress of Mathematicians, August 2018, Rio de Janeiro.

Nie, J. (2013). *Certifying Convergence of Lasserre's Hierarchy via Flat Truncation*, Math. Program. Ser. A **142**, pp. 485–510.

Nie, J. (2014). *Optimality Conditions and Finite Convergence of Lasserre's Hierarchy*, Math. Program. Ser. A **146**, pp. 97–121.

Trnovská, M. (2005). *Strong duality conditions in semidefinite programming*, J. Elec. Eng. **56**, pp. 1–5

Part I

The Moment-SOS Hierarchy for Applications in Probability and Statistics

Chapter 3

Volume and Gaussian Measure of Semi-Algebraic Sets

We describe how to apply the Moment-SOS hierarchy in order to approximate as closely as desired the Lebesgue or Gaussian measure $\mu(\mathbf{\Omega})$ of a basic semi-algebraic set $\mathbf{\Omega} \subset \mathbb{R}^n$.

3.1 Introduction

Given a basic semi-algebraic set

$$\mathbf{\Omega} = \{\, \mathbf{x} \in \mathbb{R}^n : g_j(\mathbf{x}) \geq 0, \ j = 1,\ldots,m \,\}, \tag{3.1}$$

for some polynomials $(g_j) \subset \mathbb{R}[\mathbf{x}]$, the goal is to compute (or more precisely, approximate as closely as desired) $\mu(\mathbf{\Omega})$, e.g., for the Lebesgue measure (if $\mathbf{\Omega}$ is compact) or the standard Gaussian measure on $\mathbb{R}^n$:

$$\mu(B) = \frac{1}{(2\pi)^{n/2}} \int_B \exp(-\frac{1}{2}\|\mathbf{x}\|^2)\, d\mathbf{x}, \qquad \forall B \in \mathcal{B}(\mathbb{R}^n). \tag{3.2}$$

If μ is the Lebesgue measure then we are concerned with computing the *volume* of $\mathbf{\Omega}$, a notoriously difficult problem of computational geometry. In fact as we will see, the technique that we propose also works for any measure μ for which all moments are available and satisfy some Carleman condition (only required if $\mathbf{\Omega}$ is not compact). So from now and for simplicity of exposition, we will concentrate on the Gaussian measure.

This problem is "canonical" as for a non-centered Gaussian with density $\exp(-(\mathbf{x} - \mathbf{m})^T \Sigma^{-1}(\mathbf{x} - \mathbf{m}))$ (for some real symmetric positive definite matrix Σ) one may always reduce the problem to the above one by an appropriate change of variable. Indeed after this change of variable the new domain is again a basic semi-algebraic set of the form (3.1).

Computing $\mu(\mathbf{\Omega})$ has applications in Probability & Statistics where the Gaussian measure plays a central role. With sets $\mathbf{\Omega}$ as general as in (3.1), it is a difficult and challenging problem even for rectangles $\mathbf{\Omega}$

$$\frac{1}{2\pi\sqrt{1-\rho^2}} \int_{-\infty}^{a_1} \int_{-\infty}^{a_2} \exp(-(x^2 + y^2 - 2\rho\,xy)/2(1-\rho^2))\,dx\,dy, \quad (3.3)$$

in small dimension like $n = 2$ or $n = 3$. Indeed, citing Genz (2004): *"bivariate and trivariate probability distributions computation are needed for many statistics applications ... high quality algorithms for bivariate and trivariate probability distribution computations have only more recently started to become available."*

We approximate $\mu(\mathbf{\Omega})$ as closely as desired by solving a hierarchy of semidefinite programs. Importantly, the method provides two monotone sequences $\overline{\omega}_d$ and $\underline{\omega}_d$, $d \in \mathbb{N}$, such that $\underline{\omega}_d \leq \mu(\mathbf{\Omega}) \leq \overline{\omega}_d$ for all d, and each sequence converges to $\mu(\mathbf{\Omega})$ as $d \to \infty$. Therefore at each step d the error $\overline{\omega}_d - \mu(\mathbf{\Omega})$ is bounded by $\epsilon_d := \overline{\omega}_d - \underline{\omega}_d$ and $\epsilon_d \to 0$ as $d \to \infty$. In our opinion this is an important feature which to the best of our knowledge is not present in other methods of the Literature, at least at this level of generality. It is also worth noting that the method also provides a convergent numerical scheme to approximate as closely as desired any (*à priori* fixed) number of moments of the measure $\mu_{\mathbf{\Omega}}$, the restriction of μ to $\mathbf{\Omega}$ ($\mu(\mathbf{\Omega})$ being only the mass of $\mu_{\mathbf{\Omega}}$).

3.2   An infinite-dimensional LP

We concentrate on the case where μ is the Gaussian measure in (3.2). The case where μ is the Lebesgue measure (and $\mathbf{\Omega}$ is compact) is even easier.

Recall that the goal is to compute $\mu(\mathbf{\Omega})$, where μ is the Gaussian measure in (3.2) and $\mathbf{\Omega}$ the basic semi-algebraic set in (3.1). Let $\mathbf{y} = (y_\alpha)$, $\alpha \in \mathbb{N}^n$, be the moments of μ, i.e., $y_\alpha = \int \mathbf{x}^\alpha\,d\mu$, $\alpha \in \mathbb{N}^n$, which are obtained in closed form. Call $\mu_{\mathbf{\Omega}}$ the *restriction* of μ to $\mathbf{\Omega}$, that is, $\mu_{\mathbf{\Omega}}(B) = \mu(\mathbf{\Omega} \cap B)$, for all $B \in \mathcal{B}(\mathbb{R}^n)$, and consider the new abstract problem:

$$\rho = \sup_{\phi,\nu} \left\{ \int f\,d\phi : \phi + \nu = \mu; \quad \phi \in \mathscr{M}(\Omega)_+, \ \nu \in \mathscr{M}(\mathbb{R}^n)_+ \right\}. \quad (3.4)$$

Lemma 3.1. *Let $f \in \mathbb{R}[\mathbf{x}]$ be strictly positive almost everywhere on Ω. Then $(\phi^*, \nu^*) := (\mu_\Omega, \mu - \mu_\Omega)$ is the unique optimal solution of (3.4) and $\phi^*(\Omega) = \mu(\Omega)$ while $\rho = \int_\Omega f\,d\mu$. In particular if $f = 1$ then $\rho = \mu(\Omega)$.*

Proof. As $\phi \le \mu$ and $f \ge 0$ on Ω, we have $\int f\,d\phi \le \int_\Omega f\,d\mu = \int f\,d\mu_\Omega$, and so $\theta \le \int f\,d\mu_\Omega$. On the other hand, pick $(\phi^*, \nu^*) := (\mu_\Omega, \mu - \mu_\Omega)$, to obtain $\int f\,d\phi^* = \int f\,d\mu_\Omega$, which shows that $\rho = \int f\,d\mu_\Omega$, and $\phi^*(\Omega) = \mu(\Omega)$.

Next, as $\mu + \nu = \mu$ very feasible solution is such that ϕ is absolutely continuous with respect to μ, and hence with respect to μ_Ω, denoted $\phi \ll \mu_\Omega$. In particular, by the Radon-Nikodym theorem in Ash (1972), there exists a nonnegative measurable function g on Ω such that

$$\phi(B) = \int_B g(\mathbf{x})\,d\mu_\Omega, \qquad \forall B \in \mathcal{B}(\Omega),$$

and since $\phi(B) \le \mu_\Omega(B)$ for all $B \in \mathcal{B}(\Omega)$, it follows that $g(\mathbf{x}) \le 1$, μ_Ω-a.e. on Ω. So suppose that there exists another optimal solution (ϕ, ν) with $\int f\,d\phi = \rho$, that is, $\int f(g - 1)d\mu_\Omega = 0$. As $f > 0$ μ_Ω-a.e. on Ω and $g \le 1$ on Ω, this implies $\mu_\Omega(\{\mathbf{x} : g(\mathbf{x}) < 1\}) = 0$. In other words $g = 1$, μ_Ω-a.e. on Ω which in turn implies $\phi(B) = \mu(B) = \phi^*(B)$ for all $B \in \mathcal{B}(\Omega)$. That is, $(\phi, \nu) = (\phi^*, \nu^*)$. $\square$

Problem (3.4) is an infinite-dimensional linear program on an appropriate space of measures and the following optimization problem:

$$\rho^* = \inf_{p \in \mathbb{R}[\mathbf{x}]} \left\{ \int p\,d\mu : p - f \in \mathcal{P}(\Omega); \quad p \in \mathcal{P}(\mathbb{R}^n) \right\} \quad (3.5)$$

is a *dual* of (3.4).

Indeed *weak-duality* holds because if $\phi \in \mathscr{M}(\mathbf{\Omega})_+$ (resp. $p \in \mathbb{R}[\mathbf{x}]$) is an arbitrary feasible solution of (3.4) (resp. (3.5)) then as $p \geq 0$ on $\mathbb{R}^n$, $p \geq f$ on $\mathbf{\Omega}$ and $\phi \leq \mu$, one obtains

$$\int_{\mathbb{R}^n} p \, d\mu \geq \int_{\mathbf{\Omega}} p \, d\mu \geq \int_{\mathbf{\Omega}} p \, d\phi \geq \int_{\mathbf{\Omega}} f \, d\phi,$$

and therefore $\rho^* \geq \rho$. However neither (3.4) nor the dual (3.5) are tractable.

In the next section we show how to use the Moment-SOS hierarchy to approximate ρ (and ρ^*) as closely as desired.

3.3   The Moment-SOS hierarchy

3.3.1   *Upper bounds on* $\mu(\mathbf{\Omega})$

With $d \geq d_0 := \max_j d_j$ fixed, we now consider the semidefinite program:

$$\begin{aligned}
\overline{o}_d = \sup_{\mathbf{u}, \mathbf{v} \in \mathbb{R}[\mathbf{x}]_{2d}^*} \{ L_{\mathbf{u}}(f) : \; &u_\alpha + v_\alpha = y_\alpha, \quad \forall \alpha \in \mathbb{N}_{2d}^n; \\
&\mathbf{M}_d(\mathbf{u}), \, \mathbf{M}_d(\mathbf{v}) \succeq 0 \\
&\mathbf{M}_{d-d_j}(g_j \, \mathbf{u}) \succeq 0, \quad j = 1, \ldots, m \},
\end{aligned} \tag{3.6}$$

where $\mathbf{M}_d(\mathbf{u})$ and $\mathbf{M}_{d-d_j}(g_j \, \mathbf{u})$ are respectively the moment and localizing matrices defined in §1.1. It is straightforward to see that (3.6) is a *relaxation* of (3.4), hence with $\overline{o}_d \geq \int_{\mathbf{\Omega}} f d\mu$ for all d. Indeed the constraints on $(\mathbf{u}, \mathbf{v})$ in (3.6) are only necessary for $\mathbf{u}$ and $\mathbf{v}$ to be moments of some measures ϕ on $\mathbf{\Omega}$ and ν on $\mathbb{R}^n$ respectively, such that $\phi + \nu = \mu$. The dual of (3.6) is also a semidefinite program, which in compact form reads (recalling the definition of $Q_d(g)$ in (1.5)):

$$\overline{o}_d^* = \inf_{p \in \mathbb{R}[\mathbf{x}]_{2d}} \{ \int p \, d\mu : p - f \in Q_{2d}(g); \quad p \in \Sigma[\mathbf{x}]_d \}. \tag{3.7}$$

As expected, (3.7) is a *strengthening* of (3.5). Indeed one has replaced the nonnegativity constraint $p - f \geq 0$ on $\mathbf{\Omega}$ (resp. $p \geq 0$ on $\mathbb{R}^n$) with the stronger membership $(p - f) \in Q_{2d}(g)$ (resp. the stronger membership $p \in \Sigma[\mathbf{x}]_d$).

Theorem 3.2. *Let μ be as in (3.2) and $\Omega \subset \mathbb{R}^n$ as in (3.1). Assume that both Ω and $\operatorname{supp}\mu \setminus \Omega$ have nonempty interior and let $f \in \mathbb{R}[\mathbf{x}]$ be strictly positive a.e. on Ω. Then:*

(a) Both problems (3.6) and (3.7) have an optimal solution and $\bar{o}_d = \bar{o}_d^*$ for every $d \geq d_0$.

(b) The sequence $(\bar{o}_d)$, $d \in \mathbb{N}$, is monotone non-increasing and $\bar{o}_d \to \int_\Omega f d\mu$ as $d \to \infty$. In addition if $\mathbf{u}^d$ is an optimal solution of (3.6) then $u_0^d \to \mu(\Omega)$ as $d \to \infty$.

(c) If $f = 1$ then the sequence $(\bar{o}_d)$, $d \in \mathbb{N}$, is monotone non-increasing and $\bar{o}_d \to \mu(\Omega)$ as $d \to \infty$.

Proof. (a) Recall that μ_Ω is the restriction of μ to Ω, and let $\mathbf{u} = (y_\alpha^\Omega)$, $\alpha \in \mathbb{N}_{2d}^n$, be the moments of μ_Ω, up to order $2d$. Similarly let $\mathbf{v} = (y_\alpha - u_\alpha)$, $\alpha \in \mathbb{N}_{2d}^n$, be the moments of $\mu - \mu_\Omega$. Hence $(\mathbf{u}, \mathbf{v})$ is a feasible for (3.6). In addition, as both Ω and $\operatorname{supp}\mu \setminus \Omega$ have nonempty interior, $\mathbf{M}_d(\mathbf{u}) \succ 0$ and $\mathbf{M}_d(\mathbf{v}) \succ 0$. Indeed otherwise there would exists $0 \neq p \in \mathbb{R}[\mathbf{x}]_d$ such that

$$0 = \langle \mathbf{p}, \mathbf{M}_d(\mathbf{u})\, \mathbf{p} \rangle = \int p^2 d\mu, \quad \Rightarrow p = 0 \text{ a.e. on } \Omega.$$

But then $p = 0$ since it has to vanish on some open set of Ω. The same argument also works for $\mathbf{v}$. Hence Slater's condition holds for (3.6) since $(\mathbf{u}, \mathbf{v})$ is strictly feasible solution. By a standard result in convex optimization, this in turn implies $\bar{o}_d = \bar{o}_d^*$ and moreover (3.7) has an optimal solution if $\bar{o}_d^*$ is finite. We next prove that (3.6) has an optimal solution and therefore so does (3.7). Observe that from the constraint $u_\alpha + v_\alpha = y_\alpha$ we deduce that $u_0 \leq y_0$, $v_0 \leq y_0$. Moreover all (nonnegative) diagonal elements of $\mathbf{M}_d(\mathbf{u})$ and $\mathbf{M}_d(\mathbf{v})$ are dominated by those of $\mathbf{M}_d(\mathbf{y})$. In particular:

$$\max_i [L_\mathbf{u}(x_i^{2d})] \leq \max_i [L_\mathbf{y}(x_i^{2d})]; \quad \max_i [L_\mathbf{v}(x_i^{2d})] \leq \max_i [L_\mathbf{y}(x_i^{2d})].$$

From Lemma 1.3, $|u_\alpha| \leq \tau_d$ and $|v_\alpha| \leq \tau_d$, for all $\alpha \in \mathbb{N}_{2d}^n$ (where $\tau_d := \max[y_0, \max_i [L_\mathbf{y}(x_i^{2d})]]$). Therefore the (closed) feasible of (3.6) is bounded, hence compact, which in turn implies that (3.6) has an optimal solution.

(b) That the sequence $(\bar{o}_d)$, $d \in \mathbb{N}$, is monotone non-increasing is straightforward. Next, let $(\mathbf{u}^d, \mathbf{v}^d)$ be an optimal solution of (3.6). Notice that from the proof of (a) we have seen that $|u_\alpha^d| \leq \tau_d$ and $|v_\alpha^d| \leq \tau_d$ for all $\alpha \in \mathbb{N}_{2d}^n$. Completing with zeros we now consider $\mathbf{u}^d$ and $\mathbf{v}^d$ as infinite vectors indexed by $\mathbb{N}^n$. Therefore, for each infinite sequence $\mathbf{u}^d = (u_\alpha^d)$, $\alpha \in \mathbb{N}^n$, it holds:

$$u_0^d \leq y_0 \quad \text{and} \quad 2j - 1 \leq |\alpha| \leq 2j \Rightarrow |u_\alpha^d| \leq \tau_j, \quad \forall j = 1, \ldots,$$

and similarly,

$$v_0^d \leq y_0 \quad \text{and} \quad 2j - 1 \leq |\alpha| \leq 2j \Rightarrow |v_\alpha^d| \leq \tau_j, \quad \forall j = 1, \ldots.$$

Hence by a standard argument there exists a subsequence d_k and two infinite sequences $\mathbf{u}^* = (u_\alpha^*)$ and $\mathbf{v}^* = (v_\alpha^*)$, $\alpha \in \mathbb{N}^n$, such that

$$\text{for every } \alpha \in \mathbb{N}^n, \quad u_\alpha^{d_k} \to u_\alpha^* \quad \text{and} \quad v_\alpha^{d_k} \to v_\alpha^*, \quad \text{as } k \to \infty. \tag{3.8}$$

In particular $\lim_{k \to \infty} \bar{o}_{d_k} = \lim_{k \to \infty} L_{\mathbf{u}^{d_k}}(f) = L_{\mathbf{u}^*}(f)$, and as the sequence $(\bar{o}_d)$, $d \in \mathbb{N}$, is monotone:

$$\int_\Omega f d\mu \leq \lim_{d \to \infty} \bar{o}_d = \lim_{k \to \infty} \bar{o}_{d_k} = L_{\mathbf{u}^*}(f) \quad \text{and} \quad u_\alpha^* + v_\alpha^* = y_\alpha, \tag{3.9}$$

for all $\alpha \in \mathbb{N}^n$. Then the convergence (3.8) implies that for each fixed k, $\mathbf{M}_k(\mathbf{u}^*) \succeq 0$, $\mathbf{M}_k(\mathbf{v}^*) \succeq 0$, and $\mathbf{M}_k(g_j \mathbf{u}^*) \succeq 0$ for all $j = 1, \ldots, m$. Next observe that the moment sequence $\mathbf{y}$ of the Gaussian or exponential measure μ satisfies Carleman's condition (1.11) and so μ is moment determinate. But from $L_{\mathbf{u}^*}(x_i^{2d}) \leq L_{\mathbf{y}}(x_i^{2d})$ and $L_{\mathbf{v}^*}(x_i^{2d}) \leq L_{\mathbf{y}}(x_i^{2d})$, we deduce that both $\mathbf{u}^*$ and $\mathbf{v}^*$ also satisfy Carleman's condition. This latter fact combined with $\mathbf{M}_k(\mathbf{u}^*) \succeq 0$ and $\mathbf{M}_k(\mathbf{v}^*) \succeq 0$ for all k, yields that $\mathbf{u}^*$ and $\mathbf{v}^*$ are the moment sequences of some finite Borel measures ϕ^* and ν^* on $\mathbb{R}^n$; see for instance Lasserre (2009)[Proposition 3.5].

Next, as $\mathbf{u}^*$ satisfies $L_{\mathbf{u}^*}(x_i^{2d}) \leq L_{\mathbf{y}}(x_i^{2d})$ and μ is the Gaussian or exponential measure, there is some $M > 0$ such that $L_{\mathbf{u}}(x_i^{2k}) \leq M(2k)!$ for all k. As in addition $\mathbf{M}_k(g_j \mathbf{u}^*) \succeq 0$, for all k, and all $j = 1, \ldots, m$, by Lasserre (2013)[Theorem 2.2, p. 2494] the support of ϕ^* is contained in Ω.

Hence $\phi^* \in M(\mathbf{\Omega})$ and $\nu^* \in M(\mathbb{R}^n)$. Moreover from (3.9),

$$\int \mathbf{x}^\alpha \, d(\phi^* + \nu^*) = \int \mathbf{x}^\alpha \, d\mu, \qquad \forall \alpha \in \mathbb{N}^n,$$

and as μ is moment determinate it follows that $\phi^* + \nu^* = \mu$. Hence the pair (ϕ^*, ν^*) is feasible for problem (3.4) with value $\int f d\phi^* = L_{\mathbf{u}^*}(f) \geq \int_{\mathbf{\Omega}} f d\mu = \rho$, which proves that (ϕ^*, ν^*) is an optimal solution of problem (3.4), and so $\int f d\phi^* = \int_{\mathbf{\Omega}} f d\mu$.

(c) When $f = 1$ then $\bar{o}_d = u_0^d$ and by (3.8) $u_0^{d_k} \to \mu(\mathbf{\Omega})$ as $k \to \infty$. Therefore by monotonicity of the sequence $(\bar{o}_d)$, the result follows. $\square$

Remark 3.3. (i) We emphasize that the set $\mathbf{\Omega}$ is not assumed to be compact (as in Henrion et al. (2009)).

(ii) Notice that solving (3.7) has a simple interpretation: Namely when $f = 1$ on tries to approximate the indicator function $\mathbf{x} \mapsto 1_{\mathbf{\Omega}}(\mathbf{x})$ $(= 1$ if $\mathbf{x} \in \mathbf{\Omega}$ and 0 otherwise) by polynomials of increasing degree. It is well-known that a Gibbs phenomenon occurs at points of the boundary of $\mathbf{\Omega}$. So if we choose $\mathbf{x} \mapsto f(\mathbf{x}) := \prod_{j=1}^m g_j(\mathbf{x})$ (so that f is continuous, nonnegative on $\mathbf{\Omega}$, and vanishes on the boundary of $\mathbf{\Omega}$) then the function $\mathbf{x} \mapsto f(\mathbf{x})1_{\mathbf{\Omega}}(\mathbf{x})$ is continuous, hence much easier to approximate by polynomials. For compact sets $\mathbf{\Omega}$ and the Lebesgue measure μ, it has been observed in Henrion et al. (2009) that this strategy strongly attenuates the Gibbs phenomenon and yields drastic improvements on the convergence to $\mu(\mathbf{\Omega})$.

However, when $f = 1$ the monotone convergence $\bar{o}_d \to \mu(\mathbf{\Omega})$ as $d \to \infty$ is an important attractive feature of the method because when convergence has not taken place yet, one still has the useful information that $\mu(\mathbf{\Omega}) \leq \bar{o}_d$, which is important in some applications.

We next show how (even with $f = 1$) one may still improve the convergence $\bar{o}_d \to \mu(\mathbf{\Omega})$ significantly (and thus still keep a monotone sequence of upper bounds on $\mu(\mathbf{\Omega})$). But before we show how to get a converging sequence of lower bounds on $\mu(\mathbf{\Omega})$ with same techniques.

3.3.2 *Lower bounds on* $\mu(\mathbf{\Omega})$

Let $\mathbf{\Omega}^c := \operatorname{supp} \mu \setminus \mathbf{\Omega}$ be the complement of $\mathbf{\Omega}$ in the support of μ. As μ is absolutely continuous with respect to the Lebesgue measure, the set $\mathbf{\Omega}^c$ can be written as

$$\mathbf{\Omega}^c = \cup_{\ell=1}^{s} \mathbf{\Omega}_{\ell}^c \quad \text{with} \quad \mu(\mathbf{\Omega}^c) = \sum_{\ell=1}^{s} \mu(\mathbf{\Omega}_{\ell}^c), \tag{3.10}$$

where each $\mathbf{\Omega}_{\ell}^c$ is a closed basic semi-algebraic set, $\ell = 1, \dots, s$, and the overlaps between the sets $\mathbf{\Omega}_{\ell}^c$ have μ-measure zero. (Of course the decomposition (3.10) is not unique.) For instance if μ is the Gaussian measure and $\mathbf{\Omega} = \{\mathbf{x} : g_j(\mathbf{x}) \geq 0, \, j = 1, 2\}$ then $\mathbf{\Omega}^c = \mathbf{\Omega}_1^c \cup \mathbf{\Omega}_2^c$ with:

$$\mathbf{\Omega}_1^c = \{\mathbf{x} : g_1(\mathbf{x}) \leq 0\}; \quad \mathbf{\Omega}_2^c := \{\mathbf{x} : g_1(\mathbf{x}) \geq 0; \, g_2(\mathbf{x}) \leq 0\}.$$

Then write

$$\mathbf{\Omega}_{\ell}^c = \{\mathbf{x} : g_{\ell j}(\mathbf{x}) \geq 0, \, j = 1, \dots, m_{\ell}\}, \qquad \ell = 1, \dots, s, \tag{3.11}$$

for some integer m_{ℓ} and some polynomials $(g_{\ell j}) \subset \mathbb{R}[\mathbf{x}]$, $j = 1, \dots, m_{\ell}$. Again let $d_j := \lceil (\deg g_{\ell j})/2 \rceil$ and $d_0 = \max_j d_j$.

Corollary 3.4. *Let* $\mathbf{\Omega} \subset \mathbb{R}^n$ *be as in (3.1). Assume that both* $\mathbf{\Omega}$ *and* $\mathbf{\Omega}^c$ *have nonempty interior and let* $\mathbf{\Omega}^c$ *be as in (3.10).*

(a) If $f = 1$ then for each $\ell = 1, \dots, s$, let $\overline{o}_{\ell d}^c$ be the optimal value of the semidefinite program (3.6) with $g_{\ell j}$ in lieu of g_j (and m_{ℓ} in lieu of m), and for every $d \geq d_0$, let:

$$\underline{o}_d := \mu(\mathbb{R}^n) - \left(\sum_{\ell=1}^{s} \overline{o}_{\ell d}^c \right). \tag{3.12}$$

Then the sequence $\underline{o}_d$, $d \in \mathbb{N}$, is monotone non-decreasing with $\mu(\mathbf{\Omega}) \geq \underline{o}_d$ for all $d \geq d_0$ and $\underline{o}_d \to \mu(\mathbf{\Omega})$ as $d \to \infty$.

(b) If $f \neq 1$ and f vanishes on $\partial\mathbf{\Omega}$, then for each $\ell = 1, \dots, s$, let $\overline{o}_{\ell d}^c$ be the optimal value of the semidefinite program (3.6) with f^2 and $g_{\ell j}$ in lieu of f and g_j (and m_{ℓ} in lieu of m), and for every $d \geq d_0$, let:

$$\underline{o}_d := \int f^2 d\mu - \left(\sum_{\ell=1}^{s} \overline{o}_{\ell d}^c \right). \tag{3.13}$$

Then the sequence $\underline{o}_d$, $d \in \mathbb{N}$, is monotone non-decreasing with $\int_\Omega f^2 d\mu \geq \underline{o}_d$ for all $d \geq d_0$ and $\underline{o}_d \to \int_\Omega f^2 d\mu$ as $d \to \infty$.

The proof is omitted and can be found in Lasserre (2017). As a consequence of Theorem 3.2 and Corollary 3.4, we finally obtain:

Corollary 3.5. *Let Ω be as in (3.1) and $\bar{o}_d$ be as in (3.6). If $f = 1$ then*

$$\underline{o}_d \leq \mu(\Omega) \leq \bar{o}_d, \quad \forall d \geq d_0 \quad and \quad \lim_{d \to \infty} \underline{o}_d = \mu(\Omega) = \lim_{d \to \infty} \bar{o}_d \quad (3.14)$$

where $\underline{o}_d$ is defined in (3.12). Similarly let $f \in \mathbb{R}[\mathbf{x}]$ vanish on $\partial\Omega$. If $\bar{o}_d$ in (3.6) is computed with f^2 (in lieu of f) then

$$\underline{o}_d \leq \int_\Omega f^2 d\mu \leq \bar{o}_d, \quad \forall d \geq d_0; \quad \lim_{d \to \infty} \underline{o}_d = \int_\Omega f^2 d\mu = \lim_{d \to \infty} \bar{o}_d \quad (3.15)$$

where $\underline{o}_d$ is defined in (3.13).

Why can (3.15) be also potentially interesting? Recall that any optimal solution $\mathbf{u}^d$ of (3.6) is such that $u_0^d \to \mu(\Omega)$ as $d \to \infty$. So if in (3.15) $\bar{o}_d - \underline{o}_d$ is small then it is a good indication that $u_0^d \approx \mu(\Omega)$.

3.4   Improving convergence by using Stokes' Theorem

Corollary 3.5 states nothing on the rate of convergence for the sequences $(\bar{o}_d)$ and $(\underline{o}_d)$, $d \in \mathbb{N}$, of upper and lower bounds on $\mu(\Omega)$. As already observed in Henrion et al. (2009) for compact sets Ω and the Lebesgue measure, when $f = 1$ the convergence appears to be rather slow. On the other hand, if f is positive on Ω and vanishes on $\partial\Omega$ the convergence is significantly faster but one loses the non-increasing monotonicity of the sequence $(u_0^d)_{d \in \mathbb{N}}$. Keeping the monotonicity is important because no matter how close to $\mu(\Omega)$ is $\bar{o}_d$ or $\underline{o}_d$, one still has the useful information that $\underline{o}_d \leq \mu(\Omega) \leq \bar{o}_d$ for all d (when $f = 1$).

In this section we show how to improve significantly this convergence while keeping the monotonicity of the sequences of upper and lower bounds. To do this we will use Stokes' Theorem for integration and in the sequel, to avoid technicalities, we assume that $\Omega \subset \mathbb{R}^n$ is the closure of its interior, i.e., $\Omega = \overline{\text{int}(\Omega)}$.

3.4.1   *Additional information from Stokes' Theorem*

As we already know in advance that $(\mu_\Omega, \mu - \mu_\Omega)$ is an optimal solution of the infinite-dimensional linear program (3.4), any additional information on the moments of μ_Ω will be helpful if when included as additional constraints in the relaxation (3.6), one still has a semidefinite program to solve. Fortunately for basic semi-algebraic sets Ω this is possible.

Indeed suppose for the moment that Ω is compact with smooth boundary $\partial\Omega$ and let θ_μ be the density of the Gaussian and $f \in \mathbb{R}[\mathbf{x}]$ a given polynomial. Let $e_i \in \mathbb{R}^n$ with $e_i(j) = \delta_{i=j}$, $j = 1,\ldots,n$. By Theorem 1.10, for every $i = 1,\ldots,n$, and $\alpha \in \mathbb{N}^n$:

$$\int_\Omega \frac{\partial(\mathbf{x}^\alpha f(\mathbf{x})\,\theta_\mu(\mathbf{x}))}{\partial x_i}\,d\mathbf{x} = \int_{\partial\Omega} \langle e_i, \vec{n}_\mathbf{x} \rangle\, \mathbf{x}^\alpha f(\mathbf{x})\,\theta_\mu(\mathbf{x})\,d\sigma(\mathbf{x}), \qquad (3.16)$$

where $\vec{n}_\mathbf{x}$ is the outward pointing normal at $\mathbf{x} \in \partial\Omega$ and σ is the $(n-1)$-Hausdorff measure on $\partial\Omega$. (In fact the above identity holds even if the boundary is algebraic and not smooth everywhere.) Therefore if $f \in \mathbb{R}[\mathbf{x}]$ vanishes on the boundary $\partial\Omega$ then

$$\int_\Omega \frac{\partial(\mathbf{x}^\alpha f(\mathbf{x})\,\theta_\mu(\mathbf{x}))}{\partial x_i}\,d\mathbf{x} = 0, \qquad \forall\alpha \in \mathbb{N}^n. \qquad (3.17)$$

Introduce the polynomials $p_{i,\alpha} \in \mathbb{R}[\mathbf{x}]$, $i = 1,\ldots,n$ (of degree $d_\alpha = \deg f + |\alpha| + 1$), defined by:

$$\mathbf{x} \mapsto p_{i,\alpha}(\mathbf{x}) := \frac{\partial(\mathbf{x}^\alpha f)}{\partial x_i} - \mathbf{x}^\alpha\, f(\mathbf{x})\, x_i, \qquad \forall\alpha \in \mathbb{N}^n. \qquad (3.18)$$

Then (3.17) reads

$$\int p_{i,\alpha}\,d\mu_\Omega = 0, \qquad \forall\alpha \in \mathbb{N}^n,\ i = 1,\ldots,n, \qquad (3.19)$$

and defines *linear* constraints on the moments of μ_Ω.

Extension to non-compact sets Ω. We next show that (3.17) (or, equivalently (3.19)) extends to non-compact semi-algebraic sets Ω.

Lemma 3.6. *Let $\Omega \subset \mathbb{R}^n$ be as in (3.1) (not necessarily compact) and let*

$f \in \mathbb{R}[\mathbf{x}]$ *vanish on the boundary* $\partial \Omega$. *Then with* $p_{i,\alpha} \in \mathbb{R}[\mathbf{x}]$ *as in (3.18)*,

$$\int_{\Omega} \frac{\partial (\mathbf{x}^{\alpha} f \, \exp(-\|\mathbf{x}\|^2/2))}{\partial x_i} \, d\mathbf{x} = \int_{\Omega} p_{i,\alpha} \, d\mu = 0, \qquad (3.20)$$

for every $\alpha \in \mathbb{N}^n$ *and* $i = 1, \dots, n$.

A proof can be found in Lasserre (2017).

3.4.2 *Tighter semidefinite relaxations*

So with $p_{i,\alpha} \in \mathbb{R}[\mathbf{x}]$ as in (3.18) and $r(d) := 2d - \deg f - 1$, consider the new semidefinite programs index by $d \in \mathbb{N}$:

$$\overline{\omega}_d = \sup_{\mathbf{u}, \mathbf{v} \in \mathbb{R}[\mathbf{x}]_{2d}^*} \{ u_0 : \text{ s.t. } u_{\alpha} + v_{\alpha} = y_{\alpha}, \quad \forall \alpha \in \mathbb{N}_{2d}^n;$$
$$\mathbf{M}_d(\mathbf{u}), \mathbf{M}_d(\mathbf{v}) \succeq 0$$
$$\mathbf{M}_{d-d_j}(g_j \, \mathbf{u}) \succeq 0, \quad j = 1, \dots, m, \qquad (3.21)$$
$$L_{\mathbf{u}}(p_{i,\alpha}) = 0, \qquad \forall \alpha \in \mathbb{N}_{r(d)}^n; \ i = 1, \dots, n \}.$$

(3.21) is a relaxation of the infinite-dimensional linear program (3.4) and its dual is the semidefinite program:

$$\overline{\omega}_d^* = \inf_{p \in \mathbb{R}[\mathbf{x}]_{2d}, \boldsymbol{\theta}} \{ \int p \, d\mu : \text{ s.t. } \boldsymbol{\theta} \in \mathbb{R}^{nm(d)}, \ p \in \Sigma[\mathbf{x}]_d;$$
$$p + \sum_{i=1}^{n} \sum_{\alpha \in \mathbb{N}_{m(d)}^n} \theta_{i\alpha} \, p_{i,\alpha} - 1 \in Q_{2d}(g) \}, \qquad (3.22)$$

where $m(d) = \binom{n+r(d)}{n}$ and $\boldsymbol{\theta} = (\theta_{i,\alpha}) \in \mathbb{R}^{nm(d)}$ is the vector of dual variables associated with the equality constraints $L_{\mathbf{u}}(p_{i,\alpha}) = 0$, $\alpha \in \mathbb{N}_{r(d)}^n$, $i = 1, \dots, n$.

Theorem 3.7. *Let* $\Omega \subset \mathbb{R}^n$ *be as in (3.1) and assume that both* Ω *and* $\operatorname{supp} \mu \setminus \Omega$ *have nonempty interior. Then:*

(a) Both problems (3.21) and its dual have an optimal solution and $\overline{\omega}_d = \overline{\omega}_d^*$ *for every* $d \geq \max_j d_j$.

(b) The sequence $(\overline{\omega}_d)$, $d \in \mathbb{N}$, *is monotone non-increasing and* $\overline{\omega}_d \to \mu(\Omega)$ *as* $d \to \infty$.

The proof is an almost verbatim copy of the proof of Theorem 3.2 and is omitted.

Next, exactly as we did in §3.3.2 we can provide an associated sequence of lower bounds $(\underline{\omega}_d)$, $d \in \mathbb{N}$, by considering the complement $\Omega^c := \operatorname{supp}\mu \setminus \Omega$ in $\operatorname{supp}\mu$ (with $\Omega := \Omega \cap \operatorname{supp}\mu$) and its decomposition (3.10). Then the analogue of Corollary 3.4 reads:

Corollary 3.8. *Let $\Omega \subset \mathbb{R}^n$ be as in (3.1). Assume that both Ω and Ω^c have nonempty interior and let Ω^c be as in (3.10). For each $\ell = 1, \ldots, s$, let $\overline{\omega}^c_{\ell d}$ be the optimal value of the semidefinite program (3.21) with $g_{\ell j}$ in lieu of g_j and m_ℓ in lieu of m. For every $d \geq d_0$ let:*

$$\underline{\omega}_d := \mu(\mathbb{R}^n) - \left(\sum_{\ell=1}^{s} \overline{\omega}^c_{\ell d} \right). \tag{3.23}$$

Then the sequence $\underline{\omega}_d$, $d \in \mathbb{N}$, is monotone non-decreasing with $\mu(\Omega) \geq \underline{\omega}_d$ for all $d \geq d_0$ and $\underline{\omega}_d \to \mu(\Omega)$ as $d \to \infty$.

So again but now by Theorem 3.7 and Corollary 3.8:

$$\underline{\omega}_d \leq \mu(\Omega) \leq \overline{\omega}_d, \quad \forall d \geq d_0 \quad \text{and} \quad \lim_{d \to \infty} \underline{\omega}_d = \mu(\Omega) = \lim_{d \to \infty} \overline{\omega}_d.$$

3.5   Some numerical experiments

We provide some 2D-examples to illustrate the methodology developed in §3.4. Also and importantly, for simplicity of implementation of the semidefinite programs (3.21) the moment and localizing matrices $\mathbf{M}_d(\mathbf{z})$ and $\mathbf{M}_{d-d_j}(g_j\,\mathbf{z})$ are expressed in the standard monomial basis $(\mathbf{x}^\alpha)$, $\alpha \in \mathbb{N}^n$, which is perhaps the worst choice from a numerical point of view. Indeed, the Hankel-type moment and localizing matrices matrix become rapidly ill-conditioned as their size increases. A much better choice would be the basis of Hermite polynomials that are orthogonal with respect to the Gaussian measure on $\mathbb{R}^n$ (and similarly for the exponential measure). Nonetheless the 2D-examples provided below already give some indications on the potential of the method since in many cases (but not all) good approximations $\underline{\omega}_d \leq \mu(\Omega) \leq \overline{\omega}_d$ are obtained with relatively small d (say $d = 8, 9$ or 10).

Table 3.1   Values of $\overline{w}_7$, $\underline{w}_7$ and $\epsilon_7 := 100(\overline{w}_7 - \underline{w}_7)/\underline{w}_7$ with $\sigma = .5$

	$\mathbf{u} = (0,0)$	$\mathbf{u} = (0.1, 0.1)$	$\mathbf{u} = (0.1, 0.5)$	$\mathbf{u} = (0.5, 0.5)$	$\mathbf{u} = (1,0)$
$\overline{w}_7$	0.782888	0.781978	0.783345	0.753521	0.441636
$\underline{w}_7$	0.782842	0.781932	0.783238	0.753492	0.441631
ϵ_7	0.005%	0.006%	0.01%	0.003%	0.001%

We have considered the Gaussian measures $d\mu = \exp(\|\mathbf{x}\|^2/\sigma^2)\, d\mathbf{x}$ for the three values $\sigma^2 = 0.5, 0.8, 1$. In the first set of experiments we have tested the methodology on some examples of compact sets $\Omega \subset \mathbb{R}^2$ where we could compute a very good approximate value of $\mu(\Omega)$ by other methods. For each example we have compared the hierarchy of semidefinite programs (3.6) with the hierarchy (3.21). Then we have tested the two hierarchies on some simple example of non-compact sets $\Omega \subset \mathbb{R}^2$ for which the value $\mu(\Omega)$ can be obtained exactly.

3.5.1 *On compact sets $\Omega \subset \mathbb{R}^2$*

Example 3.1. Consider non-centered Euclidean balls

$$\Omega = \{\mathbf{x} : \quad g(\mathbf{x}) \le 1\}, \quad \text{with} \quad \mathbf{x} \mapsto g(\mathbf{x}) = (x_1 - u_1)^2 + (x_2 - u_2)^2,$$

with $\mathbf{u} = (0,0)$, $(0.1, 0.1)$, $(0.5, 0.5)$, $(0.1, 0.5)$, $(1,0)$. We first show how the formulation (3.21) with $f = 1$ drastically improves convergence when comparing with the initial formulation (3.6) with $f = 1$ and $f = 1 - g$ (which vanishes on $\partial\Omega$). In Figure 3.1 the convergence is very fast for the hierarchy (3.21) (it was very slow for the hierarchy (3.6) with $f = 1$). One also observes that the convergence of the hierarchy (3.6) with $f = 1 - g$ is not monotone and not as good as (3.21). (The yellow horizontal line is the value of $\mu(\Omega)$ (exact up to 6 digits).) Iteration 1 was run with $d = 4$ (i.e. with moments up to order 8) and $d + 1, \ldots, 10$ (i.e., with moments up of order 20). With $d = 7$ (i.e. with moments up to order 14) the optimal value $s_7 = 0.573324$ obtained in (3.21) is already very close to the value 0.573132 (exact up to 6 digits).

Figure 3.2 displays the difference $\overline{o}_d - \mu(\Omega)$ for the hierarchy (3.6) with $f = 1$ and $\overline{w}_d - \mu(\Omega)$ for the hierarchy (3.21), whereas Figure 3.3 displays the respective relative errors.

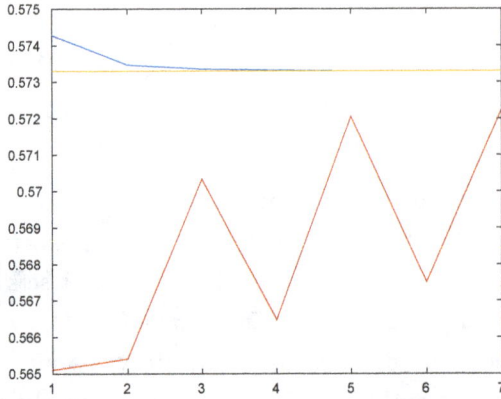

Fig. 3.1   Comparing (3.21) (with $f = 1$) and (3.6) with $f = 1$ and $f = 1 - g$

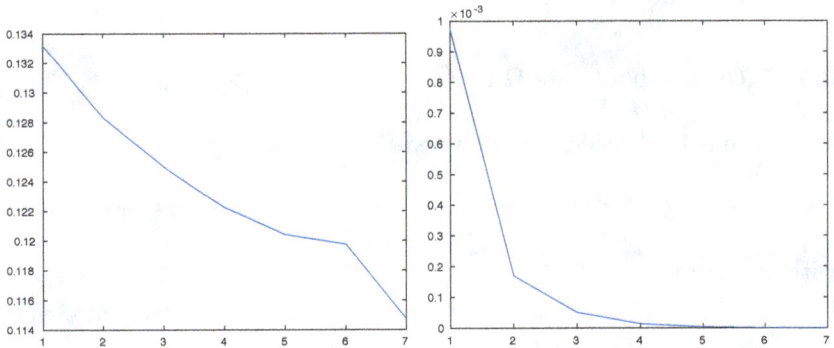

Fig. 3.2   Comparing $\rho_d - \mu(\mathbf{\Omega})$ (scheme (3.6), left) with $\overline{\omega}_d - \mu(\mathbf{\Omega})$ (scheme (3.21) right)

Table 3.2   Values of $\overline{\omega}_7$, $\underline{\omega}_7$ and $\epsilon_7 := 100(\overline{\omega}_7 - \underline{\omega}_7)/\underline{\omega}_7$ with $\sigma = .8$

	$\mathbf{u} = (0,0)$	$\mathbf{u} = (0.1, 0.1)$	$\mathbf{u} = (0.1, 0.5)$	$\mathbf{u} = (0.5, 0.5)$	$\mathbf{u} = (1,0)$
$\overline{\omega}_7$	1.906281	1.900152	1.920172	1.770550	1.183305
$\underline{\omega}_7$	1.905856	1.899708	1.919838	1.770276	1.182967
ϵ_7	0.02%	0.02%	0.017%	0.015%	0.028%

Example 3.2. Consider a set $\mathbf{\Omega} := \{\mathbf{x} \in \mathbb{R}^2 : g(\mathbf{x}) \leq 1\}$ with $\mathbf{x} \mapsto g(\mathbf{x}) := (\mathbf{x} - \mathbf{u})^T \mathbf{A} (\mathbf{x} - \mathbf{u})$ with:

$$\mathbf{A} := \begin{bmatrix} 0.4 & 0.1 \\ 0.1 & -0.4 \end{bmatrix} \begin{bmatrix} 4 & 0 \\ 0 & 8 \end{bmatrix} \begin{bmatrix} 0.4 & 0.1 \\ 0.1 & -0.4 \end{bmatrix}$$

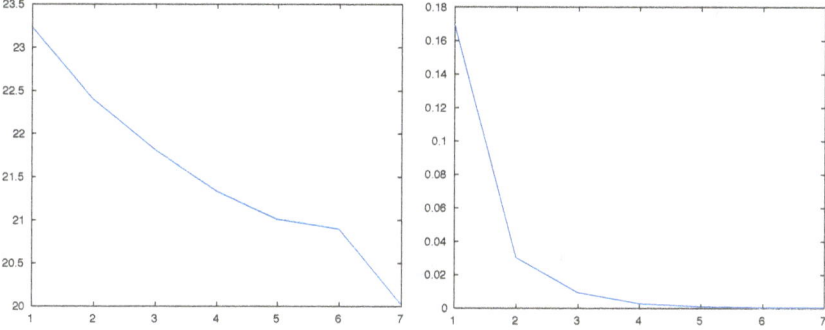

Fig. 3.3   Comparing the relative errors $100(\overline{o}_d - \mu(\mathbf{\Omega}))/\mu(\mathbf{\Omega})$ (scheme (3.6) with $f = 1$, left) with $100(\overline{w}_d - \mu(\mathbf{\Omega}))/\mu(\mathbf{\Omega})$ (scheme (3.21), right)

Table 3.3   Values of $\overline{w}_7$, $\underline{w}_7$ and $\epsilon_7 := 100(\overline{w}_7 - \underline{w}_7)/\underline{w}_7$ with $\sigma = .999$

	$\mathbf{u} = (0,0)$	$\mathbf{u} = (0.1, 0.1)$	$\mathbf{u} = (0.1, 0.5)$	$\mathbf{u} = (0.5, 0.5)$	$\mathbf{u} = (1,0)$
$\overline{w}_7$	2.811312	2.802918	2.843411	2.624901	2.947715
$\underline{w}_7$	2.806004	2.797471	2.839425	2.621340	2.946099
ϵ_7	0.19%	0.19%	0.14%	0.13%	0.05%

Table 3.4 $\mathbf{\Omega} = \{\mathbf{x} : (\mathbf{x} - \mathbf{u})^T \mathbf{A}(\mathbf{x} - \mathbf{u}) \leq 1\}$; compact case.

σ	$\overline{w}_9$	$\underline{w}_9$	$(\overline{w}_9 - \underline{w}_9)/\underline{w}_9$
		$\mathbf{u} = (0.1, 0.5)$	
1	1.728640	1.657139	4.3%
0.8	1.305168	1.298370	0.5%
0.5	0.648429	0.648415	0.002%
		$\mathbf{u} = (0.5, 0.1)$	
1	1.798363	1.730477	3.9%
0.8	1.401565	1.394053	0.53%
0.5	0.724022	0.724005	0.002%

and $\mathbf{u} = (0.1, 0.5)$, $\mathbf{u} = (0.5, 0.1)$. Results displayed in Table 3.4 show that good approximations are obtained with relatively small $d = 9$ when $\sigma \leq 1$.

3.5.2   *On non-compact sets $\mathbf{\Omega} \subset \mathbb{R}^2$*

We next provide some examples of non-compact sets $\mathbf{\Omega} \subset \mathbb{R}^2$ and show that the hierarchy (3.21) can still provide good results.

Table 3.5 $\Omega = \{\mathbf{x} : (\mathbf{x} - \mathbf{u})^T \mathbf{A}(\mathbf{x} - \mathbf{u}) \leq 1\}$;
non-compact case.

	$\mathbf{u} = (0.5, 0.1)$		
σ	$\overline{\omega}_9$	$\underline{\omega}_9$	$(\overline{\omega}_9 - \underline{\omega}_9)/\underline{\omega}_9$
1	2.829605	2.824718	0.17%
0.8	1.876731	1.876609	0.006%
	$\mathbf{u} = (0.1, 0.5)$		
σ	$\overline{\omega}_9$	$\underline{\omega}_9$	$(\overline{\omega}_9 - \underline{\omega}_9)/\underline{\omega}_9$
1	2.989832	2.986599	0.10%
0.8	1.969188	1.969103	0.004%

Table 3.6 $\Omega := \{\mathbf{x} \in \mathbb{R}^2 : x_1 + 2x_2 \geq 1\}$

σ	$\overline{\omega}_8$	$\underline{\omega}_8$	$(\overline{\omega}_8 - \underline{\omega}_8)/\underline{\omega}_8$
1.0	0.828105	0.827800	0.03%
0.8	0.4314786	0.431473	0.001%
0.5	0.080858	0.0808578	0.0003%

Example 3.3. Complement of Euclidean balls: We have first considered evaluating $\mu(\Omega^c)$ for the complement $\Omega^c := \{\mathbf{x} : \|\mathbf{x} - \mathbf{u}\|^2 \geq 1\}$ of the sets Ω considered in Example 3.1. As the lower bound $\underline{\omega}_d$ on $\mu(\Omega)$ in Example 3.1 was computed via an upper bound on $\mu(\Omega^c)$, the results in Table 3.1, Table 3.2 and Table 3.3 clearly indicate that good upper and lower bounds are also obtained for Ω^c.

Example 3.4. Non-convex quadratics: Let $\mathbf{u} = (0.1, 0.5)$, $\mathbf{u} = (0.5, 1)$,

$$\mathbf{A} := \begin{bmatrix} 0.4 & 0.1 \\ 0.1 & -0.4 \end{bmatrix} \begin{bmatrix} 4 & 0 \\ 0 & -8 \end{bmatrix} \begin{bmatrix} 0.4 & 0.1 \\ 0.1 & -0.4 \end{bmatrix},$$

and $\mathbf{x} \mapsto g(\mathbf{x}) = (\mathbf{x} - \mathbf{u})^T \mathbf{A}(\mathbf{x} - \mathbf{u})$ so that the set $\Omega := \{\mathbf{x} : g(\mathbf{x}) \leq 1\}$ is non-compact. The results displayed in Table 3.5 show that good approximations can be obtained with relatively small $d = 9$ when $\sigma \leq 1$.

Example 3.5. Half-spaces: In this example we consider the half-space $\Omega := \{\mathbf{x} \in \mathbb{R}^2 : x_1 + 2x_2 \geq 1\}$, and compute the upper and lower bounds $\overline{\omega}_d$ and $\underline{\omega}_d$ for $d = 8$. The results displayed in Table 3.6 show that good approximations can be obtained with relatively small $d = 8$ when $\sigma \leq 1$.

Example 3.6. In this example we consider the cone $\Omega := \{\mathbf{x} \in \mathbb{R}^2 : x_1 + 2x_2 \leq -0.5; x_1 \geq -0.8\}$, with apex $\mathbf{x} = (-0.8, 0.15)$, and compute the upper and lower bounds $\overline{\omega}_d$ and $\underline{\omega}_d$ for $d = 10$.

Table 3.7 $\Omega := \{\mathbf{x} \in \mathbb{R}^2 : x_1 + 2x_2 \leq -0.5;$
$x_1 \geq -0.8\}$

σ	$\overline{\omega}_{10}$	$\underline{\omega}_{10}$	$(\overline{\omega}_{10} - \underline{\omega}_{10})/\underline{\omega}_{10}$
0.5	0.208606	0.19410	7%
0.4	0.108321	0.10768	0.6%
0.3	0.041300	0.0411906	0.26%

Table 3.7 shows that when σ is relatively small then a good approximation can be obtained by using no more than $2d = 20$ moments. On the other
hand for the cone $\Omega := \{\mathbf{x} : \mathbf{x} \geq 0\}$, even with $\sigma = 0.3$ the approximation
with $d = 10$ is rather rough as we have $\overline{\omega}_{10} = 0.11172 \geq \mu(\Omega) \approx 0.070685$
and lower bound being negative is not informative. It seems that this is
because the origin (the apex of the cone Ω) is in a region where the density
$\exp(-h)$ is maximum. In this case higher order moments are needed for a
better approximation. This in turn requires some care in solving the associated semidefinite programs (3.21). Indeed, when expressed in the standard
monomial basis $(\mathbf{x}^\alpha)$, $\alpha \in \mathbb{N}$, the moment and localizing matrices in (3.21)
become ill-conditioned. An interesting issue of further investigation and be-
yond the scope of this paper is to consider alternative bases, e.g. the basis
of Hermite polynomials which are orthogonal with respect to the Gaussian
measure.

3.6 Volume of sublevel sets of polynomials

In this section $\Omega = \{\mathbf{x} : g(\mathbf{x}) \leq 1\} \subset \mathbf{B} = [-1,1]^n$ where $g \in \mathbb{R}[\mathbf{x}]_t$ is
homogeneous and nonnegative on $\mathbb{R}^n$, i.e., $g(\theta\mathbf{x}) = \theta^t g(\mathbf{x})$ for all $\theta \in \mathbb{R}$
and all $\mathbf{x} \in \mathbb{R}^n$ (so that t is even). Let μ be the Lebesgue measure on $\mathbf{B}$
and let $\#\mu$ on $[0, +\infty)$ be the pushforward measure of μ by the mapping
$g : \mathbf{B} \to \mathbb{R}$. That is:

$$\#\mu(B) := \mu(g^{-1}(B)), \qquad \forall B \in \mathcal{B}(\mathbb{R}_+). \tag{3.24}$$

In particular the moments $(\#\mu_j)_{j \in \mathbb{N}}$ of $\#\mu$, which read:

$$\#\mu_j = \int_{[0,\infty)} z^j \, d\#\mu(z) = \int_{\mathbf{B}} g(\mathbf{x})^j \, d\mu(\mathbf{x}), \qquad j = 0, 1, \dots \tag{3.25}$$

can be obtained in closed form as g is a polynomial and μ is the Lebesgue measure on the box $\mathbf{B} = [-1,1]^n$. Next, observe that $g(\mathbf{\Omega}) = \{ z \in \mathbb{R}_+ : 0 \leq z \leq 1 \}$.

Therefore:

$$\mu(\mathbf{\Omega}) = \#\mu([0,1]) = \sup_{\phi \in \mathscr{M}([0,1])_+} \{ \phi([0,1]) : \phi \leq \#\mu \}. \qquad (3.26)$$

That is, computing the n-dimensional Lebesgue volume $\mu(\mathbf{\Omega})$ reduces to computing the one-dimensional "volume" $\#\mu([0,1])$ for the measure $\#\mu$ on the real half-line $[0,\infty)$.

As all moments of $\#\mu$ are available in closed form, one may then apply the Moment-SOS hierarchy as described in §3.3, but now for a one-dimensional problem! In view of the increasing computational burden associated with the Moment-SOS hierarchy when the dimension increases, solving (3.26) should be easier than solving (3.4).

However, and as already mentioned for (3.4), the convergence is still typically slow. For solving (3.4) (with μ being the Lebesgue measure on $\mathbf{B}$) the convergence was improved by including additional moment-constraints coming from Stokes' Theorem; see §3.4. But for solving problem (3.26) one may not implement these additional Stokes' constraint because $\#\mu$ is *not* the Lebesgue measure and its density is not known explicitly.

A hierarchy of generalized eigenvalue problems

We first start with the following result:

Theorem 3.9. *The infinite-dimensional LP* (3.26) *has a unique optimal solution* $\phi^* \in \mathscr{M}([0,1])_+$ *whose moments* $\phi^* = (\phi_j^*)_{j \in \mathbb{N}}$ *satisfy:*

$$\phi_0^* = \mu(\mathbf{\Omega}); \quad \phi_j^* = \frac{n}{n + jt}\phi_0^*, \quad j = 1, 2, \ldots. \qquad (3.27)$$

Equivalently, ϕ^* *is the measure on* $[0,1]$ *with density* $\frac{n}{t} z^{n/t-1}$ *w.r.t. Lebesgue measure on* $[0,1]$.

Proof. The idea is to use Stokes' formula for an appropriate integral with vector field $X = \mathbf{x}$. Indeed, since $(1 - g^j)$ vanishes on $\partial\Omega$ for all $1 \leq j \in \mathbb{N}$:

$$
0 = \int_\Omega \mathrm{Div}(X \cdot (1 - g^j))\, d\mu = n \int_\Omega (1 - g^j)\, d\mu - \int_\Omega \langle \mathbf{x}, \nabla g^j \rangle\, d\mu
$$

$$
= n\,\mu(\Omega) - n \int_\Omega g^j\, d\mu - jt \int_\Omega g^j\, d\mu
$$

$$
= n\,\phi_0^* - (n + jt) \int_{[0,1]} z^j\, d\#\mu
$$

$$
= n\,\phi_0^* - (n + jt)\,\phi_j^*, \qquad j = 1, \ldots
$$

Then it is straightforward to verify that

$$
\frac{n}{n + jt} = \frac{n}{t} \int_0^1 z^j\, z^{n/t - 1}\, dz, \qquad j = 0, 1, \ldots
$$

$\square$

Approximating ϕ_0^*. The moment matrix $\mathbf{M}_d^*$ and localizing matrix $\mathbf{M}_{d,x(1-x)}^*$ of the measure on $[0,1]$ with density $\frac{n}{t} z^{n/t - 1}$, read:

$$
\mathbf{M}_d^*(k, \ell) = \frac{n}{n + (k + \ell - 2)\,t}; \quad k, \ell = 1, 2, \ldots \tag{3.28}
$$

$$
\mathbf{M}_{d,x(1-x)}^*(k, \ell) = \frac{n}{n + (k + \ell - 1)\,t} - \frac{n}{n + (k + \ell)\,t}, \quad k, \ell = 1, 2, \ldots
$$

From their definition, it follows easily that $\mathbf{M}_d^* \succ 0$ and $\mathbf{M}_{d,x(1-x)}^* \succ 0$ for all $d \in \mathbb{N}$, and for problem (3.26), the analogues of the semidefinite relaxations (3.21) for problem (3.4) read:

$$
\tau_d = \sup\{\tau : \tau\,\mathbf{M}_d^* \preceq \mathbf{M}_d(\#\mu)\}, \quad d = 0, 1, \ldots
$$

Equivalently $\tau_d = \lambda_{\min}(\mathbf{M}_d(\#\mu), \mathbf{M}_d^*)$ where for any two real symmetric matrices $\mathbf{A}_1, \mathbf{A}_2$, $\lambda_{\min}(\mathbf{A}_1, \mathbf{A}_2)$ denotes the smallest generalized eigenvalue of the pair $(\mathbf{A}_1, \mathbf{A}_2)$. (Notice that τ_d can be computed by a linear algebra routine with no optimization involved.) As a result we obtain:

Corollary 3.10. *Let* $\Omega = \{\mathbf{x} : g(\mathbf{x}) \leq 1\}$ *where* $g \in \mathbb{R}[\mathbf{x}]_t$ *is homogeneous of degree* t, *nonnegative on* $\mathbb{R}^n$ *and such that* $\mu(\Omega)$ *is finite. Let* $\#\mu$ *be as in* (3.25) *and* $\mathbf{M}_d^*$ *as in* (3.28). *Then:*

$$\lambda_{\min}(\mathbf{M}_d(\#\mu), \mathbf{M}_d^*) \downarrow \mu(\Omega) \quad as\ d \to \infty. \tag{3.29}$$

Example 3.7. To verify that the convergence (3.29) can be quite fast, let $n = 8$, $\mathbf{B} = [-1,1]^8$ and $\Omega := \{\mathbf{x} \in \mathbb{R}^n : \|\mathbf{x}\| \leq 1\}$. Then $\rho^* := \mu(\Omega) = \pi^{n/2}/\Gamma(1 + n/2) = 4.0587$. In Table 3.8 are displayed the values of τ_d as well as the corresponding relative error as d varies. One may see that τ_d converges to $\mu(\Omega)$ quite fast.

Extensions. Corollary 3.10 easily extends to quasi-homogeneous polynomials[1]. Next if $g \in \mathbb{R}[\mathbf{x}]_t$ is not homogeneous and $\Omega = \{\mathbf{x} : a \leq g(\mathbf{x}) \leq b\}$ is compact with $\Omega \subset (-1,1)^n$ (possibly after scaling), then write $\mathbf{x} \mapsto g(\mathbf{x}) := \sum_{k=1}^t g_k(\mathbf{x})$, for all $\mathbf{x} \in \mathbb{R}^n$, where each $g_k \in \mathbb{R}[\mathbf{x}]_k$ is homogeneous of degree k, $k = 1, \ldots, t$. Let the measure $\#\mu$ on $\mathbb{R}^t$ be the pushforward of μ on $\mathbf{B}$ by the mapping

$$\mathbf{x} \mapsto \mathbf{g}(\mathbf{x}) := \begin{bmatrix} g_1(\mathbf{x}) \\ \cdots \\ g_t(\mathbf{x}) \end{bmatrix}, \quad \mathbf{x} \in \mathbb{R}^n.$$

Proceeding as before, let $S = \mathbf{g}(\mathbf{B})$ and $\Theta := \{\mathbf{z} \in S : a \leq \sum_{k=1}^t z_k \leq b\}$. The moments $(\#\mu_\beta)_{\beta \in \mathbb{N}^t}$ of $\#\mu$ are easily obtained in closed form by:

$$\#\mu_\beta := \int_S \mathbf{z}^\beta \, d\#\mu(\mathbf{z}) = \int_\mathbf{B} g_1(\mathbf{x})^{\beta_1} \cdots g_t(\mathbf{x})^{\beta_t} \, d\mu(\mathbf{x}), \quad \forall \beta \in \mathbb{N}^t.$$

Then:

$$\mu(\Omega) = \#\mu(\Theta) = \sup_{\phi \in \mathscr{M}(\Theta)_+} \{\phi(\Theta) : \phi \leq \#\mu\}, \tag{3.30}$$

i.e., one has reduced the n-dimensional Lebesgue volume computation $\mu(\Omega)$ to the t-dimensional "volume" computation $\#\mu(\Theta)$ for the pushforward measure $\#\mu$. This is particularly interesting when $t \ll n$.

---

[1] A polynomial $p \in \mathbb{R}[\mathbf{x}]$ is quasi-homogeneous if there exists $\mathbf{u} \in \mathbb{Q}^n$ such that $p(\lambda^{u_1} x_1, \ldots, \lambda^{u_n} x_n) = \lambda p(\mathbf{x})$ for all $\lambda > 0$ and all $\mathbf{x} \in \mathbb{R}^n$.

Table 3.8 $n = 8$, $\rho^* = 4.0587$; τ_d and relative error

d	$d = 2$	$d = 3$	$d = 4$	$d = 5$	$d = 6$	$d = 7$	$d = 8$
$2^n \tau_d$	15.04	7.97	5.569	4.639	4.272	4.133	4.083
$\frac{100(2^n \tau_d - \rho^*)}{\rho^*}$	270%	96%	37%	14%	5.26%	1.83%	0.60%

However, again the convergence of the semidefinite relaxations described in §3.3 (but now associated with problem (3.30)) is typically very slow. Fortunately, one may also include additional Stokes' moments constraints that must be satisfied by the optimal solution $\phi^* \in \mathcal{M}(\Theta)_+$ of (3.30), in the spirit of §3.4. For example with $t = 2$ and vector field $X = \mathbf{x}$, consider integrals of the form

$$0 = \int_\Omega \mathrm{Div}(X \cdot h(\mathbf{x}) g_1(\mathbf{x})^i g_2(\mathbf{x})^j) \, d\mu(\mathbf{x}), \quad i, j = 0, 1, \ldots$$

where by construction the polynomial $\mathbf{x} \mapsto h(\mathbf{x}) = (b - g_1(\mathbf{x}) - g_2(\mathbf{x}))(g_1(\mathbf{x}) + g_2(\mathbf{x}) - a)$ vanishes on $\partial\Omega$. For each pair (i, j) one obtains a linear constraint on the moments $\phi^* = (\phi^*_\beta)_{\beta \in \mathbb{N}^2}$ of the unknown optimal measure ϕ^* on Θ. But in contrast to the case where g was homogeneous, this time we do not obtain an analogue of Theorem 3.9. For more details (and also other extensions), the interested reader is referred to Lasserre (2019).

3.7   Concluding remarks

We have described how to apply the Moment-SOS Hierarchy to approximate as closely as desired the Gaussian or exponential measure $\mu(\Omega)$ of a class of semi-algebraic sets $\Omega \subset \mathbb{R}^n$. It consists of solving a hierarchy of semidefinite programs of increasing size d, and of course the size of the resulting semidefinite programs increases with d. Namely the semidefinite program (3.21) has $O(n^{2d})$ variables and both (real symmetric) moment matrices $\mathbf{M}_d(\mathbf{u})$ and $\mathbf{M}_d(\mathbf{v})$ are of size $\binom{n+d}{d} \times \binom{n+d}{d}$.

So as long as one is interested in 2D- or 3D-problems, the size is not the most serious problem. However if one needs high values of d (e.g. when $\sigma \gg 1$ or $\lambda > 1$) then the precision of the SDP solvers can become a serious issue as the semidefinite programs (3.21) become ill-conditioned. In particular in such conditions the choice of the monomial basis $(\mathbf{x}^\alpha)$, $\alpha \in \mathbb{N}^n$, in which to express the moment and localizing matrices is not appropriate. To avoid ill-conditioning a more appropriate strategy is to use the basis of *Hermite polynomials* that orthogonal with respect to the Gaussian measure μ. Then one may expect to be able to solve (3.21) for higher values of d and so obtain better upper and lower bounds. In any case it is also worth mentioning that even if one is forced to stop the hierarchy (3.21) at a relatively small value of d, one has still obtained a non-trivial finite sequence of upper and lower bounds on $\mu(\mathbf{\Omega})$ for non-trivial (and possibly non-compact) sets $\mathbf{\Omega}$.

For simples sets like rectangles (3.3) and ellipsoids (with $n = 3$), in its present form this technique does not compete in terms of accuracy with *ad-hoc* procedures like in e.g. Genz (2004) (rectangles) and the recent Serra et al. (2016) (for ellipsoids). However, in view of the growing interest for semidefinite programming and its use in many applications, it is expected that more efficient packages will be available soon. For instance the semidefinite package SDPA of Fujisawa et al. (2000) has now been provided with a double precision variant[2].

Finally, it is worth mentioning a recent extension of this technique in Tacchi et al. (2019) to handle large size problems provided that there exists some structured sparsity in the description of the set $\mathbf{\Omega}$. In doing so, compact sets $\mathbf{\Omega} \subset \mathbb{R}^{100}$ can be handled relatively easily and in particular, a scaling procedure permits to avoid numerical ill-conditioning; indeed a set $\mathbf{\Omega} \subset [-1, 1]^{100}$ has extremely small volume.

3.8   Notes and sources

This chapter is essentially taken from Lasserre (2017) (which is itself an extension of Henrion et al. (2009)) and Lasserre (2019). As already mentioned, even in dimension 2 or 3, computing Gaussian measure of some simple sets can be challenging. For instance, Genz (2004) describes techniques with high accuracy results for bivariate and trivariate "rectangles"

---

[2]The SDPA-GMP, SDPA-QD and SDPA-DD versions of the standard SDPA package; see http://sdpa.sourceforge.net/family.html#sdpa.

(3.3) using sophisticated techniques to integrate Plackett's formula. Again those efficient techniques takes are very specific as they take advantage of Plackett's formula available for (3.3). Interestingly, a (complicated) formula in closed form is provided in Chandramouli and Ranganatahn (1999) via the characteristic function method.

The case of ellipsoids $\mathbf{\Omega}$ has been investigated in the pioneering work of Ruben (1962), Kotz et al. (1967a), Kotz et al. (1967b), when studying the distribution of random variables that are quadratic forms of independent normal variables. Even in small dimension it has important applications in Astronautics where for instance $\mu(\mathbf{\Omega})$ can model the probability of collision between two spatial vehicles and must be computed with very good accuracy; see for instance the works by Alfano (2005), Chan (2008), Patera (2001) and the more recent Serra et al. (2016) which combines Laplace transform techniques with the theory of D-finite functions (see e.g. Salvy (2005)). In doing so one obtains $\mu(\mathbf{\Omega})$ as a series with only nonnegative terms so that its evaluation does not involve error prone cancellations. For more details the reader is referred to Alfano (2005), Chan (2008), Patera (2001), Serra et al. (2016) and the references therein. Other applications in Astronautics and for weapon evaluation require to compute integral of bivariate Normal distributions on *convex polygons*, a case treated in Didonato et al. (1980) where at an intermediate step the authors also evaluate Gaussian integrals on (unbounded) angular regions. However the techniques developed in the above cited works and in some of the references therein, are not reproducible for more general sets $\mathbf{\Omega}$ of the form (3.1).

Section 3.6 is from Lasserre (2019). The idea of using the pushforward measure to reduce a Lebesgue volume computation in $\mathbb{R}^n$ to a "volume" computation in $\mathbb{R}$ for the pushforward measure is from Jasour et al. (2018). The use of an appropriate Stokes' formula which yields Theorem 3.9 is crucial to accelerate drastically the otherwise typically slow convergence. Several extensions (e.g. when g is not homogeneous) can be found in Lasserre (2019).

Bibliography

Alfano, S. (2005). A numerical implementation of spherical object collision probability, J. Astro. Sci. **53**(1), January–March 2005.

Anjos, M. and Lasserre, J. B. (2012). *Handbook of Semidefinite, Conic and Polynomial Optimization*, (M. Anjos and J. B. Lasserre Eds., Springer-Verlag, New York).

Ash, R. (1972). *Real Analysis and Probability*, (Academic Press Inc., San Diego).

Chan, F. K. (2008). *Spacecraft Collision Probability*, (IAA. The Aerospace Press).

Chandramouli, R., and Ranganathan, N. (1999). *Computing the Bivariate Gaussian Probability Integral*, IEEE Sign. Proc. Letters **6**, pp. 129–130.

Didonato, A. R., Jarnagin, M. P., Hageman, R. K. (1980). *Computation of the integral of the bivariate normal distribution over convex polygons*, SIAM J. Sci. Stat. Comput. **1**, pp. 179–186.

Fujisawa, K., Fukuda, M., Kojima, M., Nakata, K. (2000). *Numerical Evaluation of SDPA (Semidefinite Programming Algorithm)*, In: *High Performance Optimization*, Applied Optimization Volume **33**, Springer, New York, pp. 267–301.

Genz, A. (2004). *Numerical computation of Rectangular bivariate and trivariate Normal and t probabilities*, Statistics and Computing **14**, pp. 151–160.

Genz, A., and Monahan, J. (1999). *A Stochastic Algorithm for High Dimensional Integrals over Unbounded Regions with Gaussian Weight*, J. Comp. Appl. Math. **112**, pp. 71–81.

Henrion D., Lasserre, J. B., Lofberg, J. (2009). *Gloptipoly 3: moments, optimization and semidefinite programming*, Optim. Methods & Softwares **24**, pp. 761–779.

Henrion D., Lasserre J. B., Savorgnan C. (2009). *Approximate volume and integration for basic semialgebraic sets*, SIAM Review **51**, pp. 722–743, 2009.

Jasour, A., Hofmann, A., Williams, B. C. (2018). *Moment-Sum-Of-Squares Approach For Fast Risk Estimation In Uncertain Environments*, arXiv:1810.01577.

Kotz, S., Johnson, N. L., Boyd, D. W. (1967a). *Series representation of distribution of quadratic forms in normal variables. I. Central case*, Annals Math. Stat. **38**, pp. 823–837.

Kotz, S., Johnson, N. L., Boyd, D. W. (1967b). *Series Representations of Distributions of Quadratic Forms in Normal Variables II. Non-Central Case*, Annals Math. Stat. **38**, pp. 838–848.

Lasserre, J. B. (2013). *The K-moment problem for continuous functionals*, Trans. Amer. Math. Soc. **365**, pp. 2489–2504.

Lasserre, J. B. (2009). *Moments, Positive Polynomials and Their Applications*, Imperial College Press, London.

Lasserre, J. B. (2017). *Computing Gaussian & exponential measures of semi-algebraic sets*, Adv. Appl. Math. **91**, pp. 137–163.

Lasserre, J. B. (2019). *Volume of sublevel sets of homogeneous polynomials*, SIAM J. Appl. Alg. Geom. **3**, pp. 372–389.

Patera, R. (2001). *General method for calculating satellite conjunction probability*, J. Guidance Control & Dynamics **24**, pp. 716–722.

Ruben, H. (1962). *Probability Content of Regions Under Spherical Normal Distributions, IV: The Distribution of Homogeneous and Non-Homogeneous Quadratic Functions of Normal Variables*, Annals Math. Stat. **33**, pp. 542–570.

Salvy, B. (2005). *D-finiteness: Algorithms and applications*, In *Proceedings of the 18th International Symposium on Symbolic and Algebraic Computation*, Beijing, China, July 24–27, 2005, Manuel Kauers Editor, ACM Press, 2005, pp. 2–3.

Serra, R., Arzelier, D., Joldes, M., Lasserre, J. B., Rondepierre, A., Salvy, B. (2016). *Fast and Accurate Computation of Orbital Collision Probability for Short-Term Encounters*, J. Guidance, Control & Dynamics **39**, pp. 1–13.

Tacchi, M., Weisser, T., Lasserre, J. B., Henrion, D. (2019). *Exploiting sparsity for semi-algebraic set volume computation*. arXiv:1902.02976. Submitted.

Chapter 4

Lebesgue Decomposition of a Measure

We describe how to apply the Moment-SOS Approach in order to approximate as closely as desired moments of ν, ψ in the Lebesgue decomposition $\nu + \psi$ of a given measure μ with respect to another measure λ.

4.1 Introduction

The contribution in this chapter is in the line of research concerned with the following issue: *which type and how much of information on the support of a measure can be extracted from its moments* (a research issue outlined in a *Problem session* at the 2013 Oberwolfach meeting on *Structured Function Systems and Applications*, Charina et al. (2013)).

We are concerned with "computing" the Lebesgue decomposition $\nu + \psi$ of a measure μ with respect to another measure λ (i.e., with $\nu \ll \lambda$ and $\psi \perp \lambda$), from the sole knowledge of moments of μ and λ.

For instance and typically, μ is an atomic measure ψ (a signal) with finite support, corrupted with a small additive noise $\nu \ll \lambda$ (for some reference measure λ). Then one is interested in recovering the singular part, i.e., ψ.

The only information that we use is the sequence of moments (λ_α) and (μ_α), $\alpha \in \mathbb{N}^n$, of λ and μ. In particular, *no* à priori information on the respective supports of λ and μ is required.

We first show that computing the Lebesgue decomposition can be cast as an infinite-dimensional LP, an instance of the GMP. We then apply the

Moment-SOS hierarchy which consists of solving a hierarchy of semidefinite relaxations $(\mathbf{P}_d)$, $d \in \mathbb{N}$. The output is under the form of two finite sequences (y_α) and $(\mu_\alpha - y_\alpha)$, $\alpha \in \mathbb{N}_d^n$, whose length $\binom{n+d}{n}$ is the number of power moments up to order d. When $\nu \ll \lambda$ has a density $f \in L_\infty(\lambda)$ with $\|f\|_\infty \leq \gamma$, then the two sequences converge (as d increases) to the respective moments of ν and ψ in the Lebesgue decomposition $\nu + \psi = \mu$. Otherwise if $\|f\|_\infty > \gamma$ or if $f \notin L_\infty(\lambda)$, the two sequences converge to the respective moments of $d\nu_\gamma := (\gamma \wedge f)d\lambda$ and $\psi_\gamma := \mu - \nu_\gamma$.

4.2 Lebesgue decomposition and convex optimization

Let $\mathbf{X} \subset \mathbb{R}^n$ and let μ, λ be two finite Borel measures $\mu, \lambda \in \mathscr{M}(\mathbf{X})_+$. Let $C_\lambda := \{\phi \in \mathscr{M}(\mathbf{X})_+ : \phi \ll \lambda\}$ which is a convex cone.

4.2.1 *Lebesgue decomposition as a convex optimization problem*

Consider the infinite-dimensional optimization problem:

$$\mathbf{P}: \quad \rho = \sup_{\nu} \{\nu(\mathbf{X}) : \nu \leq \mu; \ \nu \ll \lambda; \ \nu \in \mathscr{M}(\mathbf{X})_+\}$$

$$= \sup_{\nu,\psi} \{\nu(\mathbf{X}) : \nu + \psi = \mu; \ \nu \in C_\lambda, \ \psi \in \mathscr{M}(\mathbf{X})_+\}, \quad (4.1)$$

where the notation $\nu \leq \mu$ is understood *setwise*, i.e., $\nu(B) \leq \mu(B)$ for all Borel sets $B \subset \mathbb{R}^n$.

Theorem 4.1. *The optimization problem (4.1) has a unique optimal solution $\nu^* \in \mathscr{M}(\mathbf{X})_+$ and $(\nu^*, \mu - \nu^*)$ provides the Lebesgue decomposition of μ w.r.t. λ.*

Proof.   The set $\Delta := \{\nu \in \mathscr{M}(\mathbf{X})_+ : \nu \leq \mu\} \subset \mathscr{M}(\mathbf{X})_+$ is not empty. Moreover it is bounded since $\nu(\mathbf{X}) \leq \mu(\mathbf{X})$ for all $\nu \in \Delta$. Therefore the countable additivity of ν on $\mathbf{X}$ is uniform with respect to $\nu \in \Delta$. Hence by Dunford and Schwartz (1958)[Theorem 1, p. 305] the set Δ is weakly sequentially compact. Therefore let $(\nu_n) \subset \Delta$, $n \in \mathbb{N}$, be a maximizing sequence of (4.1). By weak sequential compactness of Δ, there exists $\nu^* \in \Delta$ and a subsequence (n_k), $k \in \mathbb{N}$, such that

$$\lim_{k \to \infty} \int f \, d\nu_{n_k} = \int f \, d\nu^*, \quad \forall f \in M(\mathbf{X})^*,$$

and in particular, *setwise* convergence takes place, i.e.,

$$\lim_{k \to \infty} \nu_{n_k}(B) = \nu^*(B), \quad \forall B \in \mathcal{B}(\mathbf{X}).$$

This also implies $\rho = \lim_{k \to \infty} \nu_{n_k}(\mathbf{X}) = \nu^*(\mathbf{X})$. Next, as $\nu_{n_k} \ll \lambda$ for all k, a consequence of the above setwise convergence is that $\nu^* \ll \lambda$. It remains to prove that $\psi^* := (\mu - \nu^*) \perp \lambda$. Assume that this is not the case. Then the Lebesgue decomposition of ψ^* w.r.t. λ yields that $\psi^* = \varphi + \chi$ with $0 \neq \varphi \ll \lambda$ and $\chi \perp \lambda$. In addition, as $\nu^* + \psi^* = \mu$ we also have $\nu^* + \varphi \leq \mu$. But then $\tilde{\nu} := (\nu^* + \varphi) \in \Delta$ and $\tilde{\nu}(\mathbf{X}) = \rho + \varphi(\mathbf{X}) > \rho$, a contradiction. Therefore $\nu^* \ll \lambda$ and $\psi^* \perp \lambda$ which proves that (ν^*, ψ^*) is the Lebesgue decomposition of μ w.r.t. λ. Uniqueness of the latter implies that ν^* is the unique optimal solution of $\mathbf{P}$. □

Observe that $\mathbf{P}$ is a *convex* (but infinite dimensional) conic optimization problem, and in fact even an infinite dimensional linear programming problem (or *linear program* (LP)). A dual of $\mathbf{P}$ is the linear program

$$\mathbf{P}^*: \quad \rho^* = \inf_f \left\{ \int_{\mathbf{X}} f \, d\mu : f - 1 \in C_\lambda^*; \ f \in \mathcal{B}(\mathbf{X})_+ \right\}$$

$$= \inf_f \left\{ \int_{\mathbf{X}} f \, d\mu : f - 1 \geq 0 \quad \lambda\text{-a.e.}; \ f \in \mathcal{B}(\mathbf{X})_+ \right\}, \quad (4.2)$$

and by standard weak duality $\rho^* \geq \rho$. In fact:

Lemma 4.2. *Let (ν^*, ψ^*) be the unique optimal solution of $\mathbf{P}$, and let $B^* \in \mathcal{B}(\mathbf{X})$ be a Borel set such that $\lambda(B^*) = \lambda(\mathbf{X})$ and $\psi^*(\mathbf{X} \backslash B^*) = \psi^*(\mathbf{X})$. Then an optimal solution of $\mathbf{P}^*$ is the function $f^* \in \mathcal{B}(\mathbf{X})_+$ such that $f^*(\mathbf{x}) = 0$ on $\mathbf{X} \backslash B^*$ and $f^*(\mathbf{x}) = 1$ on B^*.*

Proof. The function f^* in Lemma 4.2 is feasible for $\mathbf{P}^*$ with associated value

$$\rho^* \leq \int_{\mathbf{X}} f^* \, d\mu = \int_{B^*} f \, d\mu = \mu(B^*) = \psi^*(B^*) + \nu^*(B^*) = \nu^*(\mathbf{X}) = \rho,$$

and the result follows since $\rho^* \geq \rho$. $\square$

So the two LPs (4.1) and (4.2) provide us with two dual characterizations of the problem. Namely:

- An optimal solution $(\nu^*, \psi^*) \in \mathscr{M}(\mathbf{X})_+^2$ of (4.1) identifies the Lebesgue decomposition of μ w.r.t. λ.
- An optimal solution $f^* \in \mathscr{B}(\mathbf{X})_+$ of (4.2) is the indicator function 1_{B^*} of the Borel set B^* on which the restriction of μ (i.e. ν^*) is absolutely continuous w.r.t. λ.

So far, to solve (4.1) we have not addressed the issue of how to handle the constraint $\nu \ll \lambda$ (or $\nu \in C_\lambda$) only from knowledge of moments of λ. Fortunately, we know how to do that if the density f^* of $\nu^* \ll \lambda$ is assumed to be in $L_\infty(\mathbf{X}, \lambda)_+$, a more restrictive case. Indeed in this case we may invoke the following result:

Theorem 4.3. *Let $\mathbf{X} \subset \mathbb{R}^n$ and let (ν_α) and (λ_α), $\alpha \in \mathbb{N}^n$, be the respective moment sequences of a finite Borel measure ν and λ on $\mathbf{X}$. Let $\mathbf{M}_d(\nu)$ and $\mathbf{M}_d(\lambda)$, be their respective moment matrices, $d = 0, 1, \ldots$. Assume that $(\lambda_\alpha)_{\alpha \in \mathbb{N}^n}$ satisfies Carleman's condition (1.11). Then the following two statements are equivalent:*
     (a) $\nu \ll \lambda$ with density $f \in L_\infty(\mathbf{X}, \lambda)_+$ and $\|f\|_\infty \leq \gamma$.
     (a) $\mathbf{M}_d(\nu) \preceq \gamma \mathbf{M}_d(\lambda)$ for every integer d and for some $\gamma > 0$.

Proof.   The implication (a) $\Rightarrow$ (b) is straightforward. Indeed, let γ be as in (a). Then:

$$\int_{\mathbf{X}} g^2 \, d\nu = \int_{\mathbf{X}} g^2 f d\lambda \leq \|f\|_\infty \int_{\mathbf{X}} g^2 d\lambda \leq \gamma \int_{\mathbf{X}} g^2 d\lambda, \qquad \forall g \in \mathbb{R}[\mathbf{x}]_d,$$

which shows that $\mathbf{M}_d(\nu) \preceq \gamma \mathbf{M}_d(\lambda)$. The reverse implication follows from Lasserre (2009)[Theorem 3.13]. $\square$

In the next section we will see that the constraint $\mathbf{M}_d(\nu) \preceq \gamma \mathbf{M}_d(\lambda)$ is easy to implement as it is a linear matrix inequality on the unknown moments of ν.

4.3   A numerical approximation scheme

In this section we show how to obtain the Lebesgue decomposition $\mu = \nu + \psi$ of μ w.r.t. λ, when the absolutely continuous part $\nu \ll \lambda$ has a density $f \in L_\infty(\mathbf{X}, \lambda)_+$ (instead of $f \in L_1(\mathbf{X}, \lambda)_+$) with $\|f\|_\infty \leq \gamma$. We also assume that both λ and μ satisfy Carleman's condition (1.11) (automatically satisfied when $\mathbf{X}$ is compact).

In fact, when $\nu \ll \lambda$ with a density $f \notin L_\infty(\mathbf{X}, \lambda)_+$ or with a density $f \in L_\infty(\mathbf{X}, \lambda)_+$ such that $\|f\|_\infty > \gamma$, the procedure that we describe below will provide in the limit, the moment sequence of the measure $\nu_\gamma \ll \lambda$ with density $f_\gamma = \gamma \wedge f$, so that $f_\gamma \in L_\infty(\mathbf{X}, \lambda)_+$ (but then of course $\psi = \mu - \nu_\gamma$ is not singular w.r.t. λ).

With $\gamma > 0$ fixed, consider the following optimization problem:

$$\rho_\gamma = \sup_\nu \{\nu(\mathbf{X}) : \nu \leq \mu; \quad \nu \leq \gamma\lambda; \quad \nu \in \mathscr{M}(\mathbf{X})_+ \}. \qquad (4.3)$$

Theorem 4.4. *Let* $(\nu^*, \mu - \nu^*)$ *be the Lebesgue decomposition of* μ *w.r.t.* λ, *and let* $f^* \in L_1(\mathbf{X}, \lambda)_+$ *be the density of* ν^*. *The Borel measure* $\nu_\gamma^* \ll \lambda$ *with density* $f_\gamma^* := \gamma \wedge f^*$ *in* $L_\infty(\mathbf{X}, \lambda)_+$ *is the unique optimal solution of (4.3).*

Proof.   Let $\nu \in \mathscr{M}(\mathbf{X})_+$ be any feasible solution of (4.3). Let B^* be a Borel set such that $\lambda(B^*) = \lambda(\mathbf{X})$ and $\psi^*(B^*) = 0$. From $\nu \leq \gamma\lambda$ we also have $\nu \ll \lambda$ and the density f_γ of ν is such that $0 \leq f_\gamma \leq \gamma$ on B^*. From $\nu \leq \mu$ we also deduce that $\nu \leq \nu^*$ on B^*, that is (as both $\nu \ll \lambda$ and $\nu^* \ll \lambda$), $f_\gamma \leq f^*$, λ-a.e. on $\mathbf{X}$, and so $f_\gamma \leq \gamma \wedge f^*$, λ-a.e. on $\mathbf{X}$. But then $\nu(\mathbf{X}) \leq \int_{\mathbf{X}} (\gamma \wedge f^*) d\lambda = \nu_\gamma^*(\mathbf{X})$. Finally, assume that there exists another optimal solution $\nu' \ll \lambda$, hence with some density $f'_\gamma \in L_\infty(\mathbf{X}, \lambda)_+$ such that $\|f'_\gamma\|_\infty \leq \gamma$. From the above argument, $f'_\gamma \leq \gamma \wedge f^*$, λ-a.e. on $\mathbf{X}$ so that

$$0 = \nu'(\mathbf{X}) - \nu_\gamma^*(\mathbf{X}) = \int_{\mathbf{X}} \underbrace{(f'_\gamma - (\gamma \wedge f^*))}_{\leq 0} d\lambda,$$

which implies that $f'_\gamma = (\gamma \wedge f^*)$, λ-a.e. on $\mathbf{X}$. This in turn implies $\nu' = \nu^*_\gamma$, the desired result. $\qquad\square$

4.3.1   The Moment-SOS hierarchy

With $\mathbf{X} \subset \mathbb{R}^n$, let μ, λ be two Borel measures on $\mathbf{X}$ of which we know all moments

$$\mu_\alpha = \int \mathbf{x}^\alpha \, d\mu; \quad \lambda_\alpha = \int \mathbf{x}^\alpha \, d\lambda, \quad \alpha \in \mathbb{N}^n.$$

Assumption 4.1. Both moment sequences (μ_α) and (λ_α), $\alpha \in \mathbb{N}$, satisfy Carleman's condition (1.11).

With $\gamma > 0$ fixed and $d \geq 1$, consider the optimization problem:

$$
\begin{aligned}
\rho_d = \sup_{\mathbf{y},\mathbf{u},\mathbf{v}} \ & \{ L_{\mathbf{y}}(1) : \\
\text{s.t. } & y_\alpha + v_\alpha = \mu_\alpha, \quad \alpha \in \mathbb{N}^n_{2d} \\
& y_\alpha + u_\alpha = \gamma \lambda_\alpha, \quad \alpha \in \mathbb{N}^n_{2d} \\
& \mathbf{M}_d(\mathbf{y}), \mathbf{M}_d(\mathbf{u}), \mathbf{M}_d(\mathbf{v}) \succeq 0 \},
\end{aligned}
\tag{4.4}
$$

where $L_{\mathbf{y}}$ (resp. $\mathbf{M}_d(\mathbf{y})$) is the Riesz functional (resp. the moment matrix) associated with the sequence $\mathbf{y}$; see §1.1.

Problem (4.4) is a semidefinite program. It is straightforward to check that (4.4) is a relaxation of (4.3) and so $\rho_d \geq \rho_\gamma$ for every $d \in \mathbb{N}$. Its dual is also a semidefinite program which reads:

$$
\begin{aligned}
\rho^*_d = \inf_{p,q,\sigma} \ & \int p \, d\mu + \gamma \int q \, d\lambda \\
\text{s.t.} \quad & p + q - 1 = \sigma; \quad p, q, \sigma \in \Sigma[\mathbf{x}]_d.
\end{aligned}
\tag{4.5}
$$

Theorem 4.5. *The semidefinite program (4.4) has an optimal solution* $(\mathbf{y}^*, \mathbf{u}^*, \mathbf{v}^*)$ *and there is no duality gap between (4.4) and its dual (4.5), i.e., $\rho_d = \rho^*_d$.*

Proof.   First observe that (4.4) has the trivial solution $\mathbf{y} = 0$ and $(v_\alpha) =$

$(\mu_\alpha), (u_\alpha) = \gamma(\lambda_\alpha)$. From the moment constraints we immediately have:

$$L_{\mathbf{y}}(x_i^{2d}) \leq \int \mathbf{x}_i^{2d} d\lambda; \quad L_{\mathbf{v}}(x_i^{2d}) \leq \int \mathbf{x}_i^{2d} d\mu; \quad L_{\mathbf{u}}(x_i^{2d}) \leq \gamma \int \mathbf{x}_i^{2d} d\lambda,$$

for all $i = 1, \ldots, n$. In addition $L_{\mathbf{y}}(1) \leq \mu_0$, $L_{\mathbf{v}}(1) \leq \mu_0$ and $L_{\mathbf{u}}(1) \leq \gamma\lambda_0$. Let

$$\tau_1 := \max[\mu_0, \max_i[\int \mathbf{x}_i^{2d} d\mu]]; \quad \tau_2 := \gamma \max[\lambda_0, \max_i[\int \mathbf{x}_i^{2d} d\lambda]]. \quad (4.6)$$

By Lemma 1.3 it follows that

$$\sup_{\alpha \in \mathbb{N}_{2d}^n} |y_\alpha| \leq \tau_1; \quad \sup_{\alpha \in \mathbb{N}_{2d}^n} |v_\alpha| \leq \tau_1; \quad \sup_{\alpha \in \mathbb{N}_{2d}^n} |u_\alpha| \leq \tau_2. \quad (4.7)$$

Therefore the feasible set of (4.4) is compact and so there exists an optimal solution $(\mathbf{y}^*, \mathbf{u}^*, \mathbf{v}^*)$. In addition, it also follows that the set of optimal solutions of (4.4) is also compact. Therefore there is no duality gap between (4.4) and its dual (4.5), that is, $\rho_d = \rho_d^*$; see for instance Trnovská (2005). $\square$

Theorem 4.6. *Let Assumption 4.1 hold. For every $d \geq 1$, let $(\mathbf{y}^d, \mathbf{v}^d, \mathbf{u}^d)$ be an arbitrary optimal solution of (4.4) and by completing with zeros, consider $\mathbf{y}^d, \mathbf{u}^d$ and $\mathbf{v}^d$ as elements of $\mathbb{R}[\mathbf{x}]^*$.*

Then the sequence of triplets $(\mathbf{y}^d, \mathbf{v}^d, \mathbf{u}^d)_{d \in \mathbb{N}} \subset (\mathbb{R}[\mathbf{x}]^*)^3$ converges to $(\mathbf{y}^*, \mathbf{v}^*, \mathbf{u}^*) \in (\mathbb{R}[\mathbf{x}]^*)^3$ as $d \to \infty$, that is, for every fixed $\alpha \in \mathbb{N}^n$:

$$\lim_{d \to \infty} y_\alpha^d = y_\alpha^*; \quad \lim_{d \to \infty} v_\alpha^d = \mu_\alpha - y_\alpha^*; \quad \lim_{d \to \infty} u_\alpha^d = \gamma\lambda_\alpha - y_\alpha^*. \quad (4.8)$$

Moreover, $\mathbf{y}^*$ is the vector of moments of the measure $\nu_\gamma^* \leq \mu$, unique optimal solution of (4.3), with density $f_\gamma^* = (\gamma \wedge f^*) \in L_\infty(\mathbf{X}, \lambda)$ and $\|f_\gamma^*\|_\infty \leq \gamma$. Similarly $\mathbf{v}^*$ is the vector of moments of the measure $\psi^* := \mu - \nu_\gamma^*$.

Proof. Let $(\mathbf{y}^d, \mathbf{v}^d, \mathbf{u}^d) \in (\mathbb{R}[\mathbf{x}]^*)^3$, $d \in \mathbb{N}$, be as in Theorem 4.6 and define the triplet $(\hat{\mathbf{y}}^d, \hat{\mathbf{v}}^d, \hat{\mathbf{u}}^d) \in (\mathbb{R}[\mathbf{x}]^*)^3$, $d \in \mathbb{N}$, by:

$$\hat{y}_\alpha^d = y_\alpha^d / \tau_{1d}, \quad \forall \alpha : 2d - 1 \leq |\alpha| \leq 2d$$

$$\hat{v}_\alpha^d = v_\alpha^d / \tau_{1d}, \quad \forall \alpha : 2d - 1 \leq |\alpha| \leq 2d$$

$$\hat{u}_\alpha^d = u_\alpha^d / \tau_{2d}, \quad \forall \alpha : 2d - 1 \leq |\alpha| \leq 2d,$$

for all $d = 1, 2, \ldots$, where τ_{1d}, τ_{2d} are defined in (4.7). Therefore

$$\sup_{\alpha \in \mathbb{N}^n} |\hat{y}_\alpha^d| \leq 1; \quad \sup_{\alpha \in \mathbb{N}^n} |\hat{v}_\alpha^d| \leq 1; \quad \sup_{\alpha \in \mathbb{N}^n} |\hat{u}_\alpha^d| \leq 1,$$

and the sequence $\hat{\mathbf{y}}^d$, $d \in \mathbb{N}$, (as well as sequences $\hat{\mathbf{v}}^d$ and $\hat{\mathbf{u}}^d$, $d \in \mathbb{N}$) is contained in the unit ball of ℓ_∞ (compact and sequentially compact for the weak-$*$ topology $\sigma(\ell_\infty, \ell_1)$). By Banach-Alaoglu Theorem (see e.g. Dunford and Schwartz (1958)) there exist a subsequence (d_k) and sequences $\hat{\mathbf{y}}^*$, $\hat{\mathbf{v}}^*$ and $\hat{\mathbf{u}}^*$ (each in the unit ball of ℓ_∞) such that for each $\alpha \in \mathbb{N}^n$:

$$\lim_{k \to \infty} \hat{y}_\alpha^{d_k} = \hat{y}_\alpha^*; \quad \lim_{k \to \infty} \hat{v}_\alpha^{d_k} = \hat{v}_\alpha^*; \quad \lim_{k \to \infty} \hat{u}_\alpha^{d_k} = \hat{u}_\alpha^*.$$

Therefore,

$$\lim_{k \to \infty} y_\alpha^{d_k} = y_\alpha^*; \quad \lim_{k \to \infty} v_\alpha^{d_k} = v_\alpha^*; \quad \lim_{k \to \infty} u_\alpha^{d_k} = u_\alpha^*, \qquad (4.9)$$

with:

$$y_\alpha^* = \hat{y}_\alpha^* \cdot \tau_{1d}, \quad \forall \alpha : 2d - 1 \leq |\alpha| \leq 2d$$
$$v_\alpha^* = \hat{v}_\alpha^* \cdot \tau_{1d}, \quad \forall \alpha : 2d - 1 \leq |\alpha| \leq 2d$$
$$u_\alpha^* = \hat{u}_\alpha^* \cdot \tau_{2d}, \quad \forall \alpha : 2d - 1 \leq |\alpha| \leq 2d,$$

for all $d = 1, 2, \ldots$. Next, fix $d \in \mathbb{N}$, arbitrary. From (4.9)

$$0 \preceq \mathbf{M}_d(\mathbf{y}^*) \preceq \mathbf{M}_d(\mu); \quad 0 \preceq \mathbf{M}_d(\mathbf{v}^*) \preceq \mathbf{M}_d(\mu); \quad 0 \preceq \mathbf{M}_d(\mathbf{u}^*) \preceq \gamma \mathbf{M}_d(\lambda).$$

In particular:

$$L_{\mathbf{y}^*}(x_i^{2k}) \leq \int_{\mathbf{X}} x_i^{2k} \, d\mu; \quad L_{\mathbf{v}^*}(x_i^{2k}) \leq \int_{\mathbf{X}} x_i^{2k} \, d\mu; \quad L_{\mathbf{u}^*}(x_i^{2k}) \leq \gamma \int_{\mathbf{X}} x_i^{2k} \, d\lambda.$$
$$(4.10)$$

Recall that by Assumption 4.1 Carleman's condition (1.11) holds for μ and λ. Therefore (4.10) implies that Carleman's condition also holds for $\mathbf{y}^*$, $\mathbf{v}^*$, and $\mathbf{u}^*$. Next, as $\mathbf{M}_d(\mathbf{y}^*) \succeq 0$, $\mathbf{M}_d(\mathbf{v}^*) \succeq 0$ and $\mathbf{M}_d(\mathbf{u}^*) \succeq 0$, then by Lasserre (2009)[Proposition 3.5], $\mathbf{y}^*$, $\mathbf{v}^*$, and $\mathbf{u}^*$ are the respective moment sequences of finite Borel measures ν, ψ, and ϕ on $\mathbf{X}$. In addition, ν, ψ, and ϕ are moment determinate and since (4.8) holds it follows that

$$\nu + \psi = \mu; \quad \nu + \phi = \gamma \lambda,$$

which shows that ν is a feasible solution of (4.3). We also have

$$\rho_\gamma \leq \lim_{k\to\infty} \rho_{d_k} = \lim_{k\to\infty} L_{\mathbf{y}^{d_k}}(1) = L_{\mathbf{y}^*}(1) = \nu(\mathbf{X}),$$

which proves that ν is an optimal solution of (4.3). But by Theorem 4.4, the optimal solution of (4.3) is unique. Therefore all accumulation points of $(\mathbf{y}^d, \mathbf{v}^d, \mathbf{u}^d)$, $d \in \mathbb{N}$, are identical since they are the moment sequences of ν_γ^*, $\mu - \nu_\gamma^*$ and $\gamma\lambda - \nu_\gamma^*$, respectively, that is, (4.8) holds. $\qquad\square$

The meaning of Theorem 4.6 is as follows: Recall that $\nu^* + \psi^*$ (with $\nu^* \ll \lambda$ and $\psi^* \perp \lambda$) is the (unique) Lebesgue decomposition of μ. Then:

- Either ν^* has a density $f \in L_\infty(\mathbf{X}, \lambda)_+$ with $\|f\|_\infty \leq \gamma$ in which case in the limit one obtains all moments of ν^* and ψ^*, or

- ν^* does not have a density $f^* \in L_\infty(\mathbf{X}, \lambda)_+$ with $\|f^*\|_\infty \leq \gamma$. In this case, μ can be decomposed into a sum $\nu_1 + \nu_2$ where ν_1 has a density $\gamma \wedge f^* \in L_\infty(\mathbf{X}, \lambda)_+$ and $\nu_2 = \mu - \nu_1 \in \mathscr{M}(\mathbf{X})_+$. In the limit one obtains all moments of ν_1 and ν_2 (but $\nu_2 \not\perp \lambda$, i.e. ν_2 is not singular w.r.t. λ).

4.3.2 *Recovering an (atomic) singular part*

A case of particular interest is when the singular part ψ^* ($\perp \lambda$) of the Lebesgue decomposition $\nu^* + \psi^*$ of μ w.r.t. λ, is atomic and supported on finitely many atoms. Assume that ν^* has a density in $L_\infty(\mathbf{X}, \lambda)_+$ with $\|f^*\|_\infty \leq \gamma$. Then by Theorem 4.6,

$$\lim_{d\to\infty} v_\alpha^d = \mu_\alpha - y_\alpha^* = \int_{\mathbf{X}} \mathbf{x}^\alpha \, d\psi^*, \qquad \forall \alpha \in \mathbb{N}^n, \tag{4.11}$$

where $(\mathbf{y}^d, \mathbf{v}^d, \mathbf{u}^d)$ is an optimal solution of the semidefinite program (4.4).

Next, if ψ^* has a finite support, say m points $\mathbf{x}_1, \ldots, \mathbf{x}_m \in \mathbf{X}$, its moment matrix $\mathbf{M}_d(\mathbf{v}^*)$ has (finite) rank m for all $d \geq d_0$, for some d_0. In particular, $\operatorname{rank} \mathbf{M}_{d_0+1}(\mathbf{v}^*) = \operatorname{rank} \mathbf{M}_{d_0}(\mathbf{v}^*) = m$. By Curto and Fialkow (2005) this property is indeed a certificate that (v_α^*), $\alpha \in \mathbb{N}_{d_0+1}^n$ is the truncated moment sequence of a measure supported on $m = \operatorname{rank} \mathbf{M}_{d_0}(\mathbf{v}^*)$ points of $\mathbb{R}^n$. There is even a linear algebra procedure to extract the m points $\mathbf{x}_1, \ldots, \mathbf{x}_m$ from the sole knowledge of the finitely many moments (v_α^*), $|\alpha| \leq d_0 + 1$; see e.g. Henrion and Lasserre (2005).

Proposition 4.7. *Let $\{\mathbf{x}_1, \ldots, \mathbf{x}_m\} \subset \mathbf{X}$ be the support of the singular part ψ^* in the Lebesgue decomposition $\nu^* + \psi^*$ of μ w.r.t. λ. Let d_0 be such that $\operatorname{rank} \mathbf{M}_d(\mathbf{v}^*) = m$ for all $d \geq d_0$, and let $\eta > 0$ be the smallest strictly positive eigenvalue of $\mathbf{M}_{d_0}(\mathbf{v}^*)$. Let $\mathbf{v}^d$ be part of an optimal solution of (4.4).*

Then for every fixed $\epsilon > 0$, there exists $d_\epsilon > d_0$ such that the first respective $s(d_0) - m$ and $s(d_0 + 1) - m$ eigenvalues (arranged in increasing order) of $\mathbf{M}_{d_0}(\mathbf{v}^d)$ and $\mathbf{M}_{d_0+1}(\mathbf{v}^d)$ are less than ϵ and their last respective m eigenvalues are larger than $\eta/2$.

Proof. Recall that $s_d = \binom{n+d}{n}$ is the size of the moment matrix $\mathbf{M}_d(\mathbf{v}^*)$. Let η be the smallest strictly positive eigenvalue of $\mathbf{M}_{d_0}(\mathbf{v}^*)$. The eigenvalues of $\mathbf{M}_{d_0}(\cdot)$ and $\mathbf{M}_{d_0+1}(\cdot)$ (arranged in increasing order) are continuous functions of the entries and (4.11) holds. So by (4.11), given $\epsilon > 0$ there exists $d_\epsilon > 0$ such that for every $d \geq d_\epsilon$, the moment matrices $\mathbf{M}_{d_0}(\mathbf{y}^d)$ and $\mathbf{M}_{d_0+1}(\mathbf{y}^d)$ have m strictly positive eigenvalues with value larger than $\eta/2$ while their other respective $s_{d_0} - m$ and $s_{d_0+1} - m$ eigenvalues have value smaller than ϵ. $\qquad\square$

A practical procedure. So one may propose the following numerical procedure with an *à priori* fixed integer $p > 0$.

- A threshold 10^{-p} is proposed to "declare" zero an eigenvalue of $\mathbf{M}_d(\mathbf{v}^d)$ as follows. Compute the eigenvalues of $\mathbf{M}_d(\mathbf{v}^d)$ and check whether they can be grouped into two disjoint sets A and B such that

$$\sigma \in B \Rightarrow \frac{\sigma}{\theta_A} < 10^{-p}, \quad \text{with } \theta_A = \arg\min\{\sigma : \sigma \in A\}.$$

  In view of (4.11) this eventually happens when d is sufficiently large and with $\#A = m$. (However it may happen earlier and with $\#A \neq m$.) So once one has found such sets A and B then one considers that the rank of $\mathbf{M}_d(\mathbf{y}^d)$ is $\#A$.
- Once an optimal solution $(\mathbf{y}^d, \mathbf{v}^d, \mathbf{u}^d)$ has been computed, check whether there is some $k \leq d - 1$ such that $\operatorname{rank} \mathbf{M}_k(\mathbf{v}^d) = \operatorname{rank} \mathbf{M}_{k+1}(\mathbf{v}^d)$ where "rank" has the above numerical meaning.

- If the above rank-condition holds one considers that $\mathbf{M}_k(\mathbf{v}^d)$ is the truncated moment sequence of a measure supported on $t := \operatorname{rank} \mathbf{M}_k(\mathbf{v}^d)$ points of $\mathbb{R}^n$. The extraction procedure in Henrion and Lasserre (2005) can be applied and yields t points $\mathbf{x}_1, \ldots, \mathbf{x}_t$.

Of course the rank condition $\operatorname{rank} \mathbf{M}_k(\mathbf{v}^d) = \operatorname{rank} \mathbf{M}_{k+1}(\mathbf{v}^d)$ (with $k \leq d - 1$) can happen earlier than when $d \geq d_0$ (recall that d_0 is not known in advance). In this case there is no guarantee that the extracted points are indeed the support of ψ^*.

4.3.3 *Some numerical experiments*

Given two measures μ, λ on $\mathbf{X} \subset \mathbb{R}^n$ and their respective moment sequences (μ_α) and (λ_α), $\alpha \in \mathbb{N}^n$, with no loss of generality we may and will assume that μ is a probability measure (otherwise replace μ_α with μ_α / μ_0 for all $\alpha \in \mathbb{N}^n$).

Example 4.1. The first example is one-dimensional with one atom for the singular part. Let $\mathbf{X} = [0, 1]$, $a, b, c \in \mathbf{X}$, $a < b$, and let λ be the Lebesgue measure on $\mathbf{X}$. Let ν_{ab} be the probability measure distributed uniformly on $[a, b] \subset \mathbf{X}$ and

$$\mu = \underbrace{p\,\nu_{ab}}_{\nu^*} + \underbrace{(1 - p)\,\delta_c}_{\psi^*},$$

where δ_c is the Dirac measure at the point c and $p \in (0, 1)$ is some fixed scalar. With $a = 0.1$, $b = 0.7$, $c = 0.4$, and $\gamma = 2p$, one solves (4.4) with $d = 9$ (hence overall we look at moments up to order 18), to obtain an optimal solution $(\mathbf{y}^d, \mathbf{v}^d, \mathbf{u}^d)$. The first 5 (normalized) moments of ψ^* read

$$\begin{bmatrix} 1.00000 \ 0.40000 \ 0.16000 \ 0.06400 \ 0.02560 \end{bmatrix}$$

while the first 5 (normalized) moments of ν^* read

$$\begin{bmatrix} 1.00000 \ 0.40000 \ 0.19000 \ 0.10000 \ 0.05602 \end{bmatrix}.$$

In Table 4.1 are displayed the first 5 "moments" v_k^d / v_0^d, $k = 0, \ldots 4$, computed in (4.4) and the resulting relative errors with those of ψ^*, for different

Table 4.1   Example 4.1: First 5 approximate moments of ψ^* (normalized)

$p = 0.1$				
1.00000	0.3998	0.1601	0.0642	0.0258
0%	0.05%	0.06%	0.31%	0.76%
$p = 0.2$				
1.00000	0.39936	0.16009	0.06435	0.02597
0%	0.15%	0.05%	0.55%	1.45%
$p = 0.3$				
1.00000	0.39861	0.15982	0.06434	0.02606
0%	0.34%	0.11%	0.54%	1.79%
$p = 0.4$				
1.00000	0.39785	0.15979	0.06462	0.02639
0%	0.53%	0.12%	0.96%	2.9%
$p = 0.5$				
1.00000	0.39662	0.15956	0.06484	0.02672
0%	0.85%	0.27%	1.29%	4.2%
$p = 0.6$				
1.00000	0.39481	0.15937	0.06534	0.02736
0%	1.31%	0.39%	2.05%	6.4%

values of $p \in (0,1)$. Similarly in Table 4.2 are displayed the first 5 "moments" y_k^d/y_0^d, $k = 0, \ldots, 4$, computed in (4.4) and the resulting relative errors with those of ν^* (normalized). As one may expect, the quality of the approximation is very good for small p and the slightly deteriorates when p increases.

Recall that by Theorem 4.5 the optimal value ρ_d of (4.4) is such that $\rho_d \to \rho_\gamma$ as $d \to \infty$. However, it is worth noting that ρ_d is not very close to $\rho_\gamma = p$ when $d \leq 10$. So it seems that the semidefinite hierarchy (4.4), $d \in \mathbb{N}$, succeeds well in identifying relatively fast the support of ψ^* and ν^*, but not so well to obtain their respective masses p and $1 - p$.

Example 4.2. The second example is also one-dimensional but with two atoms for the singular part. Let $0 < p < 1$, $\mathbf{X} = [0,1]$, $a, b, c_1, c_2 \in \mathbf{X}$, $a < b$, and let λ be the Lebesgue measure on $\mathbf{X}$. Let ν_{ab} be the probability measure distributed uniformly on $[a,b] \subset \mathbf{X}$ and

$$\mu = \underbrace{p\,\nu_{ab}}_{\nu^*} + \underbrace{\frac{(1-p)}{2}\left(\delta_{c_1} + \delta_{c_2}\right)}_{\psi^*}.$$

With $a = 0.1$, $b = 0.7$, $c_1 = 0.4$, $c_2 = 0.5$, and $\gamma = 2p$, one solves (4.4)

Table 4.2   Example 4.1: First 5 approximate moments of ν^* (normalized)

$p = 0.1$				
1.00000	0.4018	0.1872	0.0959	0.05241
0%	0.45%	1.5%	4.2%	6.8%
$p = 0.2$				
1.00000	0.40232	0.18775	0.09640	0.05269
0%	0.58%	1.2%	3.7%	6.3%
$p = 0.3$				
1.00000	0.40294	0.18849	0.09699	0.05311
0%	0.73%	0.79%	3.09%	5.46%
$p = 0.4$				
1.00000	0.40288	0.18837	0.09688	0.05303
0%	0.71%	0.86%	3.21%	5.62%
$p = 0.5$				
1.00000	0.40294	0.18848	0.09698	0.05311
0%	0.73%	0.8%	3.10%	5.47%
$p = 0.6$				
1.00000	0.40291	0.18846	0.09698	0.05311
0%	0.72%	0.81%	3.11%	5.46%

with $d = 9$ (hence overall we look at moments up to order 18). The first 5 moments of the measure $(\delta_{c_1} + \delta_{c_2})/2$ read

$$\left[1.00000 \ 0.45000 \ 0.20500 \ 0.09450 \ 0.04405 \right].$$

In Table 4.3 one displays the first 5 approximate "moments" (v_k^d/v_0^d), $k = 0, \ldots, 4$, and the resulting relative errors with those of ψ^*, for various values of $p \in (0, 1)$. Similarly, in Table 4.4 one displays the first 5 approximate "moments" (y_k^d/y_0^d), $k = 0, \ldots, 4$, and the resulting relative errors with those of ν^*. Again one observes that the support of ψ^* is relatively well recovered with few moments ($d \leq 10$). However, and as in Example 4.1, the optimal value ρ_d is not very close to ρ_γ when $d \leq 10$.

Example 4.3. The third example is two-dimensional with the singular part ψ^* being uniformly supported on the point $\mathbf{x} = (1, 2)$. Then with $p \in (0, 1)$,

$$\mu = \underbrace{p \nu^* + (1 - p) \delta_{(1,2)}}_{\psi^*}; \qquad \nu^* \ll \lambda,$$

where ν^* is the (normalized) Gaussian measure with density $\exp(-x_1^2 - x_2^2)$, $\lambda = 2\nu^*$ and $\gamma = 2p$. With $d = 9$, results are displayed in Table 4.5 for

Table 4.3 Example 4.2: First 5 approximate "moments" of ψ^* (normalized)

$p = 0.1$				
1.00000	0.44952	0.20435	0.09390	0.04359
0%	0.11%	0.31%	0.62%	1.04%
$p = 0.2$				
1.00000	0.44934	0.20404	0.09358	0.04332
0%	0.14%	0.46%	0.96%	1.63%
$p = 0.3$				
1.00000	0.44910	0.20354	0.09305	0.04288
0%	0.19%	0.71%	1.53%	2.64%
$p = 0.4$				
1.00000	0.44896	0.20305	0.09247	0.04239
0%	0.23%	0.94%	2.14%	3.76%
$p = 0.5$				
1.00000	0.44894	0.20250	0.091741	0.04173
0%	0.23%	1.2%	2.91%	5.24%
$p = 0.6$				
1.00000	0.44916	0.20175	0.09063	0.04071
0%	0.18%	1.58%	4.09%	7.5%

Table 4.4 Example 4.2: First 5 approximate "moments" of ν^* (normalized)

$p = 0.1$				
1.00000	0.41114	0.19720	0.10378	0.05781
0%	2.78%	3.79%	3.78%	3.20%
$p = 0.2$				
1.00000	0.40892	0.19520	0.10230	0.05680
0%	2.23%	2.74%	2.30%	1.39%
$p = 0.3$				
1.00000	0.40844	0.19477	0.10198	0.05659
0%	2.11%	2.51%	1.98%	1.02%
$p = 0.4$				
1.00000	0.40789	0.19427	0.10162	0.05635
0%	1.97%	2.24%	1.62%	0.59%
$p = 0.5$				
1.00000	0.40742	0.19384	0.10131	0.05614
0%	1.85%	2.02%	1.31%	0.2%
$p = 0.6$				
1.00000	0.40693	0.19342	0.10101	0.05594
0%	1.73%	1.8%	1.01%	0.13%

different values of the weight $p \in (0,1)$. The first line displays the maximum relative error between the computed moment vector $\mathbf{v}^d/v_0^d$ and the moments of $\psi^* = \delta_{(1,2)}$ up to order 4. The second line displays the maximum relative

Table 4.5   Example 4.3: Relative error on the "moments" up to order 4 of ψ^* and ν^* (normalized)

	$p = 0.1$	$p = 0.2$	$p = 0.3$	$p = 0.4$	$p = 0.5$	$p = 0.6$	$p = 0.7$	$p = 0.8$
ψ^*	0.02%	0.05%	0.08%	0.11%	0.22%	0.30%	0.41%	0.63%
ν^*	0.02%	0.05%	0.03%	0.04%	0.09%	0.04%	0.01%	0.06%
ρ_9	0.1005	0.2010	0.3014	0.4019	0.5026	0.6028	0.7033	0.8035

Table 4.6   Example 4.4: Maximum relative error on the "moments" up to order 4 of ψ^* and ν^* (normalized)

	$p = 0.1$	$p = 0.2$	$p = 0.3$	$p = 0.4$	$p = 0.5$	$p = 0.6$	$p = 0.7$	$p = 0.8$
ψ^*	0.03%	0.06%	0.08%	0.14%	0.20%	0.26%	0.44%	0.71%
ν^*	1.45%	1.92%	2.15%	2.11%	2.14%	2.24%	2.21%	2.23%
ρ_9	0.1012	0.2019	0.3028	0.4035	0.5040	0.6047	0.7056	0.8062

error between the (normalized) second-order moments $\int x_1^2 d\nu^*$ and $\int x_2^2 d\nu^*$ and their approximation in the vector $\mathbf{y}^d/y_0^d$ (the first-order moments and $\int x_1 x_2 d\nu^*$ vanish while their approximation in $\mathbf{y}^d/y_0^d$ are less than 0.008). The third line displays ρ_d that ideally should be close to p.

Example 4.4. The fourth example is two-dimensional with the singular part ψ^* being uniformly supported on the two points $(1, 2)$ and $(-2, 1)$. Then with $p \in (0, 1)$,

$$\mu = p\nu^* + \underbrace{(1 - p)\left(\delta_{(1,2)} + \delta_{(-2,1)}\right)/2}_{\psi^*}; \qquad \nu^* \ll \lambda,$$

where ν^* is the (normalized) Gaussian measure with density $\exp(-x_1^2 - x_2^2)$, $\lambda = \nu^*$ and $\gamma = 2p$. With $d = 9$, results are displayed in Table 4.6 for different values of the weight $p \in (0, 1)$. The first line displays the maximum relative error between the computed moment vector $\mathbf{v}^d/v_0^d$ and the moments of $\psi^* = \delta_{(1,2)}$ up to order 4. The second line displays the maximum relative error between the (normalized) second-order moments $\int x_1^2 d\nu^*$ and $\int x_2^2 d\nu^*$ and their approximation in the vector $\mathbf{y}^d/y_0^d$ (the first-order moments and $\int x_1 x_2 d\nu^*$ vanish while their approximation in $\mathbf{y}^d/y_0^d$ are less than 0.011). Again the third line displays ρ_d that ideally should be close to p.

Example 4.5. The fifth example is two-dimensional with the singular part ψ^* being uniformly supported on the unit circle $\{\mathbf{x} \in \mathbb{R}^2 : x_1^2 + x_2^2 = 1\}$.

Then with $p \in (0, 1)$,

$$\mu = p\nu^* + (1 - p)\psi^*; \qquad \nu^* \ll \lambda,$$

where ν^* is the normalized Gaussian measure with density $\exp(-x_1^2 - x_2^2)$, $\lambda = \nu^*$ and $\gamma = 2p$. With $d = 7$ Table 4.7 displays the relative error between the moments $\int x_1^2 d\psi^*$, $\int x_1^4 d\psi^*$ and $\int x_1^2 x_2^2 d\psi^*$ (the odd moments being zero) and their respective approximation in $\mathbf{v}^d/v_0^d$, for the value of $p = 0.1$, 0.2, 0.3 and 0.4. For larger values of d the semidefinite solver encounters numerical difficulties and the numerical output cannot be trusted. The last column also displays $L_{\mathbf{v}^d/v_0^d}((x_1^2 + x_2^2 - 1)^2)$ which ideally should be $\int (x_1^2 + x_2^2 - 1)^2) d\psi^*/\psi^*(\mathbb{R}^2) = 0$ since for every $\alpha \in \mathbb{N}^n$, $v_\alpha^d \to \int \mathbf{x}^\alpha d\psi^*$ as $d \to \infty$; recall (4.11) when $\mathbf{v}^d$ is a moment sequence.

Concerning the approximation of the moments of ν^*: The relative error between the second-order moment $\int x_1^2 d\nu^*$ (normalized) and its approximation (in the computed vector of moments $\mathbf{y}^d/y_0^d$) is less than 1% for all $p = 0, 1$, 0.2, 0.3 and 0.4. On the other hand, for the order-4 moments $\int x_1^4 d\nu^*$ and $\int x_1^2 x_2^2 d\nu^*$ (normalized) the relative error is about 20%.

In Table 4.8 the same results are displayed but this time when ν^* is uniformly distributed on the unit box $[-1, 1]^2$, $\lambda = \nu^*$ and $\gamma = 2p$. In this case the relative error between the second-order moment $\int x_1^2 d\nu^*$ (normalized) and its approximation (in the computed vector of moments $\mathbf{y}^d/y_0^d$) is about 11% for all $p = 0, 1$, 0.2, 0.3 and 0.4. For the order-4 moment $\int x_1^4 d\nu^*$ (normalized) it is about 13% in all cases, and for the order-4 moment $\int x_1^2 x_2^2 d\nu^*$ (normalized) it is about 8% in all cases.

So in both Example 4.4 and Example 4.5 the singular part ψ^* (normalized) is well recovered even though the absolutely continuous part ν^* is not so well recovered (in particular the moments of order 4 are not very accurate).

As illustrated in Example 4.5, numerical problems can be encountered relatively fast (e.g. when $d > 7$ in Example 4.5). In the next section we briefly discuss how to (partly) address some numerical issues.

Table 4.7   Example 4.5: $d = 7$; ν^* Gaussian; relative error on the "moments" $\int x_1^2 d\psi^*$, $\int x_1^4 d\psi^*$ and $\int x_1^2 x_2^2 d\psi^*$ of ψ^* (normalized)

	x_1^2	x_1^4	$x_1^2 x_2^2$	$L_{\mathbf{v}^d/v_0^d}((x_1^2 + x_2^2 - 1)^2)$
$p = 0.1$	0.19%	0.52%	0.53%	0.001
$p = 0.2$	0.47%	1.28%	1.28%	0.003
$p = 0.3$	0.94%	2.76%	2.76%	0.009
$p = 0.4$	1.87%	5.93%	5.93%	0.02

Table 4.8   Example 4.5: $d = 7$; ν^* uniform on the unit box; relative error on the "moments" $\int x_1^2 d\psi^*$, $\int x_1^4 d\psi^*$ and $\int x_1^2 x_2^2 d\psi^*$ of ψ^* (normalized)

	x_1^2	x_1^4	$x_1^2 x_2^2$	$L_{\mathbf{v}^d}((x_1^2 + x_2^2 - 1)^2)$
$p = 0.1$	0.26%	0.93%	0.61%	0.002
$p = 0.2$	0.62%	2.22%	1.47%	0.0004
$p = 0.3$	1.15%	4.09%	2.76%	0.0008
$p = 0.4$	1.87%	6.97%	5.27%	0.0016

4.4   Concluding remarks

In the above simple examples one has encountered some numerical difficulties when $d > 10$ (i.e. when moments of order > 20 appear) and even when $d > 7$ for Example 4.5. Such numerical problems were not due to the size but rather to the ill-conditioning of the semidefinite program (4.4) (in turn due to the use of the monomial basis $(\mathbf{x}^\alpha)_{\alpha \in \mathbb{N}^n}$). Clearly the basis of monomials $(\mathbf{x}^\alpha)$, $\alpha \in \mathbb{N}$, mainly used for modeling and algorithm implantation convenience, is a very bad choice from a numerical viewpoint. This is especially true as we use semidefinite solvers for which such issues can be crucial even for relatively small size matrices, as observed in the examples when $d > 10$. However other (and better choices) of basis are possible. For instance let $(\mathscr{L}_\alpha)$, $\alpha \in \mathbb{N}^n$, be the basis of polynomials orthonormal w.r.t. λ. They can be obtained from the moments $(\lambda_\alpha)_{\alpha \in \mathbb{N}^n}$ by simple computation of certain determinants, as described in e.g. Dunkl and Xu (2001) and Helton et al. (2008); see e.g. the discussion in Lasserre (2016).

Finally, we have not treated the problem in full generality as one obtains the required information on the Lebesgue decomposition (ν, ψ) only when the density of ν w.r.t. λ is in $L_\infty(\lambda)$ with norm bounded by γ, fixed à priori. Otherwise one obtains a partial information only. However, so far and to

the best of our knowledge, this is the first systematic numerical scheme at
this level of generality.

4.5   Notes and sources

This chapter is essentially from Lasserre (2016) and as already mentioned,
it is in the line of research about extraction of information on the support
of a measure from knowledge of its moments.

For instance, the old and classical L-moment problem asks for moment
conditions to ensure that the underlying unknown measure μ is absolutely
continuous with respect to some reference measure ν, and with a density
in $L_\infty(\nu)$. See for instance Diaconis and Freedman (2004), Putinar (1996),
Putinar (1998), more recently Lasserre (2013b), and the many references
therein. Other works are concerned with recovering information about the
support of a measure μ from knowledge of its moments and some *à priori*
information on its support. For instance, in Lasserre (2011), bounds on the
support of μ are provided from knowledge of moments of its marginals. In
Collowald et al. (2017) and Gravin et al. (2012), the support is assumed
to be a convex polytope $P \subset \mathbb{R}^n$ and the vertices of P can be recovered
from finitely many directional moments of the Lebesgue measure on P.
Similarly, in Lasserre (2013a) and Lasserre and Putinar (2015) one may
recover the boundary of a semi-algebraic set $S \subset \mathbb{R}^n$ from moments of the
Lebesgue measure on S. In super-resolution one is interested in recovering
the discrete support (a finite set of points) of a signal from finitely many
moments of its associated signed measure; see the ground-breaking work of
de Castro and Gamboa (2012) and Candès and Fernández-Granda (2014),
and its recent multivariate extension in de Castro et al. (2017). In Lasserre
(2015) one is interested in detecting from knowledge of moments whether a
given measure $d\mu(x,t)$ on $[0,1]^2$ with given marginal dt on $[0,1]$ is supported
on a curve $(t, x(t)) \subset [0,1]^2$ for some measurable function $x : [0,1] \rightarrow [0,1]$.

Bibliography

Candès, E. J. and Fernández-Granda C. (2014). *Towards a mathematical theory of super-resolution*, Comm. Pure Appl. Math. **67**, pp. 906–956.

Charina, M., Lasserre, J. B., Putinar, M., Stöckler, J. (2013). *Structured Function Systems and Applications*, Oberwolfach Workshop, 24th February-March 2nd, Oberwolfach Report No. 11/2013, pp. 579–655, DOI: 10.4171/OWR/2013/11, 2013.

Collowald, M., Cuyt, A., Hubert, E., Lee Wen-Shi, Celis, O. S. (2017). *Numerical reconstruction of convex polytopes from directional moments*, Adv. Comput. Math., to appear.

Curto, R. E. and Fialkow, L. A. (2005). *The truncated K-moment problem in several variables*, J. Operator Theory **54**, pp. 189–226.

de Castro, Y., Gamboa, F., Henrion, D., Lasserre, J. B. (2017). *Exact solutions to Super Resolution on semi-algebraic domains in higher dimensions*, IEEE Trans. Info. Theory **63**, pp. 621–630.

de Castro, Y., and Gamboa, F. (2012). *Exact reconstruction using Beurling minimal extrapolation*, J. Math. Anal. Appl. **395**, pp. 336–354.

Diaconis, P. and Freedman, D. (2004). *The Markov moment problem and de Finetti's Theorem: Part I*, Math. Z. **247**, pp. 183–199.

Dunford, N. and Schwartz, J. (1958). *Linear Operators. Part I: General Theory*, John Wiley & Sons, Inc., New York, USA.

Dunkl, C. F. and Xu, Y. (2001). *Orthogonal Polynomials of Several Variables*, Cambridge University Press, Cambridge, UK.

Gravin, N., Lasserre, J. B., Pasechnik, D., Robins, S. (2012). *The Inverse*

Moment Problem for Convex Polytopes, Discrete & Comp. Geom. **48**, pp. 596–621.

Helton, J. W., Lasserre, J. B., Putinar, M. (2008). *Measures with zeros in the inverse of their moment matrix*, Annals Prob. **36**, pp. 1453–1471.

Henrion, D. and Lasserre, J. B. (2005). *Detecting global optimality and extracting solutions in GloptiPoly*, in *Positive Polynomials in Control*, (D. Henrion and A. Garulli Eds.), Lecture Notes on Control and Information Sciences, Vol. 312, Springer Verlag, Berlin, pp. 293–310.

Lasserre, J. B. (2016). *Lebesgue decomposition in action via semidefinite relaxations*, Adv. Comput. Math. **42**, pp. 1129–1148.

Lasserre, J. B. (2011). *Bounding the support a measure from its marginal moments*, Proc. Amer. Math. Soc. **139**, pp. 3375–3382.

Lasserre, J. B. (2013a). *Recovering an homogeneous polynomials from moments of its levels set*, Discrete & Comput. Geom. **50**, pp. 673–678.

Lasserre, J. B. and Netzer, T. (2007). *SOS approximations of nonnegative polynomials via simple high degree perturbations*, Math. Z. **256**, pp. 99–112.

Lasserre, J. B. and Putinar, M. (2015). *Algebraic-exponential data recovery from moments*, Discrete & Comput. Geometry **54**, pp. 993–1012.

Lasserre, J. B. (2015). *Moments and Legendre-Fourier Series for Measures supported on Curves*, SIGMA **11** No 077, 10 pages.

Lasserre, J. B. (2013b). *Borel measures with a density on a compact semi-algebraic set*, Archiv. der Mathematik **101**, pp. 361–371.

Lasserre, J. B. (2009). *Moments, Positive Polynomials and Their Applications*, Imperial College Press, London.

Putinar, M. (1996). *Extremal solutions of the two-dimensional L-problem of moments, I*, J. Funct. Anal. **136**, pp. 331–364.

Putinar, M. (1998). *Extremal solutions of the two-dimensional L-problems of moments, II*, J. Approx. Theory **92**, pp. 38–58.

Trnovská, M. (2005). *Strong duality conditions in semidefinite programming*, J. Elec. Eng. **56**, pp. 1–5.

Chapter 5

Super Resolution on Semi-Algebraic Domains

We describe how to apply the Moment-SOS hierarchy to solve Super Resolution problems in Signal Processing.

5.1 Introduction

The early formulation of the Super Resolution problem can be identified as the ability of faithfully reconstruct a high-dimensional sparse vector from the observation of a low-pass filter. This situation models important applications in imaging spectroscopy Harris et al. (1994), image processing Park et al. (2003), radar imaging Odendaal et al. (1994), or astronomy Makovoz and Marleau (2005). Briefly it boils down to solving

$$\min_{\mathbf{x}} \{ \|\mathbf{x}\|_1 : \mathbf{A}\,\mathbf{x} = \mathbf{b} \}, \quad \mathbf{b} = (b_k) \in \mathbb{R}^m, \quad \mathbf{A} \in \mathbb{R}^{m \times N}, \tag{5.1}$$

where $\mathbf{b} = \mathbf{A}\mathbf{x}^*$ for some unknown "sparse" vector $\mathbf{x}^*$, and $m \ll N$.

If the number s of non-zero entries of $\mathbf{x}^*$ is small and the matrix $\mathbf{A}$ enjoys some geometric property, namely the *Null Space Property* that depends only on its kernel and s, then one can exactly reconstruct $\mathbf{x}^*$ by solving (5.1). The term "Super Resolution" appears in these early works because $\mathbf{A}$ is related to the discretization of some Fourier transform. This success story enlighted the sparsity-inducing property of minimizing the ℓ_1-norm and initiated the *Compressed Sensing Theory*.

A more general and abstract framework is as follows: Suppose that $\phi \in \mathscr{M}(\mathbf{X})$ is a *signed* atomic measure on $\mathbf{X} \subset \mathbb{R}^n$, supported on finitely many atoms $\mathbf{z}^*(i) \in \mathbf{X}$, $i = 1, \ldots, s$, that is,

$$\phi = \sum_{i=1}^{s} x_i^* \, \delta_{\mathbf{z}^*(i)}, \quad 0 \neq x_i^* \in \mathbb{R}.$$

The issue is: How to recover the support $\mathbf{z}^*(i) \in \mathbf{X}$ and the weights $x_i^* \in \mathbb{R}$, $i = 1, \ldots, s$, from the sole knowledge of a few (finitely many) moments $b_\alpha \, (= \int_{\mathbf{X}} a_\alpha(\mathbf{x}) \, d\phi(\mathbf{x}))$, $\alpha \in \Gamma$, of ϕ only?

In general $\mathbf{x} \mapsto a_\alpha(\mathbf{x})$ are the natural power moments $\mathbf{x}^\alpha$, and $\Gamma \subset \mathbb{N}^n$ is a finite set of "small" cardinality.

The strategy is: (i) Consider the optimization problem:

$$\inf_{\phi \in \mathscr{M}(\mathbf{X})} \left\{ \|\phi\|_{TV} : \int_{\mathbf{X}} a_\alpha \, d\phi = b_\alpha, \quad \alpha \in \Gamma \right\}, \qquad (5.2)$$

where $\| \cdot \|_{TV}$ denotes the total variation norm of measures, and (ii) the Moment-SOS hierarchy is an appropriate tool for solving (5.2).

A rationale behind (5.2). Why is (5.2) a natural model to solve our recovery problem? To fix ideas consider the univariate case with $\mathbf{X} = [0, 1] \subset \mathbb{R}$. Observe that $\|\phi\|_{TV} = \sum_{i=1}^{s} |x_i^*|$, the ℓ_1-norm $\|\mathbf{x}\|_1$ of the vector of coefficients of the weights of ϕ. Next, the constraints $\int_{\mathbf{x}} a_\alpha \, d\mu = b_\alpha$, $\alpha \in \Gamma$, read:

$$\sum_{i=1}^{s} x_i^* \, z^*(i)^k = b_k, \quad k = 0, \ldots, m.$$

So suppose that the s unknown points $(z^*(i))_{i=1,\ldots,s}$ of the support of ϕ belong to a grid $(z(i))_{i=1,\ldots,N}$ of say N equidistant points in $[0, 1]$ with N large. Then (5.2) is equivalent to solving (5.1) with $\mathbf{A}_{kj} = (z(j))^k$.

So observe that in the formulation (5.2) we do not need know an à priori grid $(z(j))_{j=1,\ldots,N} \subset \mathbf{X}$. The problem is stated directly on an appropriate space of finite signed measures on $\mathbf{X}$.

Theorem 5.1 below states in an informal manner a seminal contribution of Candès and Fernandez-Granda (2014) for $\mathbf{X} = \mathbb{T} := \{\, z \in \mathbb{C} : z\bar{z} = 1\,\}$ (the torus) and $\mathbf{X} = \mathbb{T}^2$ (the bi-torus). It turns out that the result also holds for $\mathbb{T}^n$, $n \in \mathbb{N}$, as proved in Josz et al. (2019); see also Chapter 6 of this book.

> **Theorem 5.1.** *Let ϕ be an atomic signed measure on $\mathbf{X}$ with finitely many atoms. If the atoms of ϕ are sufficiently "geometrically separated" and there are sufficiently many moment conditions, then ϕ is the unique optimal solution of (5.2).*

5.2 Super resolution as an instance of GMP

We here assume that in (5.2) $\Gamma = \mathbb{N}_t^n$ for some t, and $a_\alpha = \mathbf{x}^\alpha$, $\alpha \in \mathbb{N}_t^n$. To model Problem (5.2) as an instance of the GMP, the signed measure $\phi \in \mathscr{M}(\mathbf{X})$ is replaced with its Jordan decomposition $\phi^+ - \phi^-$, where $\phi^+, \phi^- \in \mathscr{M}_+(\mathbf{X})$, to obtain

$$\inf_{\phi^+, \phi^- \in \mathscr{M}_+(\mathbf{X})} \left\{ \int_{\mathbf{X}} d(\phi^+ + \phi^-) : \int_{\mathbf{X}} \mathbf{x}^\alpha \, d(\phi^+ - \phi^-) = b_\alpha, \quad \alpha \in \mathbb{N}_t^n \right\}, \quad (5.3)$$

which is an instance of the GMP; see Definition 1.5.

Lemma 5.2. *If (ϕ^+, ϕ^-) is an optimal solution of (5.3) then $\int_{\mathbf{X}} d(\phi^+ + \phi^-) = \|\phi^+ - \phi^-\|_{TV}$ and $\phi := \phi^+ - \phi^-$ is an optimal solution of (5.2).*

Proof. Let (ϕ^+, ϕ^-) be an optimal solution of (5.3). Let $\phi := \phi^+ - \phi^-$ and assume that ϕ^+ and ϕ^- are not mutually singular. Let $\nu^+ - \nu^-$ be the Jordan decomposition of ϕ so that $\int_{\mathbf{X}} d(\nu^+ + \nu^-) = \|\phi\|_{TV}$ and

$$\int_{\mathbf{X}} d(\phi^+ + \phi^-) > \int_{\mathbf{X}} d(\nu^+ + \nu^-) = \|\phi\|_{TV}.$$

Moreover as $b_\alpha = \int_{\mathbf{X}} \mathbf{x}^\alpha d\phi = \int_{\mathbf{X}} \mathbf{x}^\alpha d(\nu^+ - \nu^-)$, it follows that (ν^+, ν^-) is also a feasible solution of (5.3) but with a better value than (ϕ^+, ϕ^-), a contradiction. Therefore ϕ^+ and ϕ^- are mutually singular and the result follows.   □

For every $p \in \mathbb{R}[\mathbf{x}]_t$, let $\|p\|_\infty := \sup_{\mathbf{x} \in \mathbf{X}} |p(\mathbf{x})|$. The following optimization problem

$$\sup_{p \in \mathbb{R}[\mathbf{x}]_t} \left\{ \sum_{\alpha \in \mathbb{N}_t^n} p_\alpha\, b_\alpha : \|p\|_\infty \leq 1 \right\} \qquad (5.4)$$

is a dual of (5.3) as weak-duality holds. Indeed let (ϕ^+, ϕ^-) and p be two feasible solutions of (5.3) and (5.4) respectively. Then as $\mathbf{X}$ is compact,

$$\sum_\alpha p_\alpha\, b_\alpha = \int_{\mathbf{X}} p\, d(\phi^+ - \phi^-) \leq \|p\| \int_{\mathbf{X}} d(\phi^+ + \phi^-) = \int_{\mathbf{X}} d(\phi^+ + \phi^-),$$

the desired result. In fact *strong duality* holds.

Theorem 5.3. *Let* $\mathbf{X}$ *be compact with nonempty interior. Then there is no duality gap between the LP (5.3) and its dual (5.4) and both have an optimal solution.*

Proof.   Recall that $s(t) = \binom{n+t}{n}$. Define $r(\phi^+, \phi^-) \in \mathbb{R}^{s(t)+1}$ by:

$$r(\phi^+, \phi^-) := \left[\int_{\mathbf{X}} d(\phi^+ + \phi^-), \int_{\mathbf{X}} \mathbf{x}^\alpha\, d(\phi^+ - \phi^-)_{\alpha \in \mathbb{N}_t^n} \right]$$

and the set

$$R := \{ r(\phi^+, \phi^-) : \phi^+, \phi^- \in \mathscr{M}(\mathbf{X})_+ \} \subset \mathbb{R}^{s(t)+1}.$$

Let ρ and ρ^* be the respective optimal values of (5.3) and (5.4). By Barvinok (2002)[Theorem 7.2], $\rho = \rho^*$ provided that ρ is finite and R is closed. Finiteness of ρ follows from existence of a feasible solution and nonnegativity of the objective function $\int_{\mathbf{X}} d(\phi^+ + \phi^-)$. To prove that R is closed we have to show that for any sequence $(\phi_n^+, \phi_n^-)_{n \in \mathbb{N}} \in \mathscr{M}(\mathbf{X})_+ \times \mathscr{M}(\mathbf{X})_+$ such that $r(\phi_n^+, \phi_n^-) \to r \in \mathbb{R}^{s(d)+1}$ as $n \to \infty$, one has $r = r(\phi, \psi)$ for some finite measures $\phi, \psi \in \mathscr{M}(\mathbf{X})_+$. Since the supports of all measures are contained in a compact set, and since $\int_{\mathbf{X}} d(\phi_n^+ + \phi_n^-) \to r_0$ all measures ϕ_n^+, ϕ_n^- are norm-bounded, uniformly in n. Therefore, from the weak-* compactness (and weak-* sequential compactness) of the unit ball (Banach-Alaoglu's Theorem) of $\mathcal{M}(\mathbf{X})_+$, there is a subsequence $(\phi_{n_k}^+, \phi_{n_k}^-)_{k\in\mathbb{N}}$ that converges weakly-* to an element $(\phi, \psi) \in \mathcal{M}(\mathbf{X})_+ \times \mathcal{M}(\mathbf{X})_+$. In particular, for every $\alpha \in \mathbb{N}^n$:

$$\lim_{k\to\infty} \int_{\mathbf{X}} \mathbf{x}^\alpha \, d\phi_n^+ = \int_{\mathbf{X}} \mathbf{x}^\alpha \, d\phi; \quad \lim_{k\to\infty} \int_{\mathbf{X}} \mathbf{x}^\alpha \, d\phi_n^- = \int_{\mathbf{X}} \mathbf{x}^\alpha \, d\psi,$$

which in turn implies $\lim_{k\to\infty} r(\phi_{n_k}^+, \phi_{n_k}^-) = r(\phi, \psi)$. This proves that R is closed and so $\rho = \rho^*$. From the above we also conclude that (5.3) has an optimal solution. It remains to show that (5.4) also has an optimal solution. We may identify $p \in \mathbb{R}[\mathbf{x}]_t$ with its vector of coefficients. Let $\mathbf{p} \in \mathbb{R}^{s(t)}$ be the vector of coefficients of $p \in \mathbb{R}[\mathbf{x}]_t$, and let $\mathbf{v}_t(\mathbf{x})$ be the vector of monomials up to degree t. The set

$$U := \{\, \mathbf{p} \in \mathbb{R}^{s(t)} : \|\mathbf{v}_t(\mathbf{x})^T \mathbf{p}\|_\infty \leq 1\,\} \tag{5.5}$$

of feasible solutions of (5.4) is a closed convex subset of a finite-dimensional Euclidean space, and it contains the origin. Since the objective function of (5.4) is continuous on U, the optimum is attained if U is bounded. Suppose that U is not bounded. Then there exists a sequence $(\mathbf{p}_n)_{n\in\mathbb{N}} \subset U$ such that $\|\mathbf{p}_n\| \to \infty$ as $n \to \infty$. Write $\mathbf{p}_n = \lambda_n \mathbf{q}_n$, with $\|\mathbf{q}_n\| = 1$. Notice that $0 < \lambda_n \to \infty$ and $\mathbf{q}_n \in U$ because $0 \in U$ and U is convex. Then

$$\|\mathbf{v}_t(\mathbf{x})^T \mathbf{p}_n\|_\infty = \|\mathbf{v}_t(\mathbf{x})^T \lambda_n \mathbf{q}_n\|_\infty = \lambda_n \|\mathbf{v}_t(\mathbf{x})^T \mathbf{q}_n\|_\infty \leq 1,$$

so that $\|\mathbf{v}_t(\mathbf{x})^T \mathbf{q}_n\|_\infty \leq \lambda_n^{-1} \to 0$ as $n \to \infty$. Since $\|\mathbf{q}_n\| = 1$, there exists a subsequence n_k and $\mathbf{q}$ with $\|\mathbf{q}\| = 1$ such that $\mathbf{q}_{n_k} \to \mathbf{q}$ as $k \to \infty$ and $\|\mathbf{v}_d(\mathbf{x})^T \mathbf{q}\|_\infty = 0$. As $\mathbf{X}$ has nonempty interior, this implies that $\mathbf{q} = 0$, a contradiction. $\qquad\square$

5.2.1 The Moment-SOS hierarchy

Let $\mathbf{X} \subset \mathbb{R}^n$ be the compact basic semi-algebraic set

$$\mathbf{X} := \{\mathbf{x} \in \mathbb{R}^n : g_j(\mathbf{x}) \geq 0, \, j = 1, \ldots, m\}, \tag{5.6}$$

and with no loss of generality, we may and will assume that (possibly after scaling) $g_1(\mathbf{x}) = 1 - \|\mathbf{x}\|^2$. Also define $\mathbf{x} \mapsto g_0(\mathbf{x}) := 1$ for all $\mathbf{x}$ and let $d_j := \lceil (\deg g_j)/2 \rceil$ for all $j = 0, \ldots, m$, and $\bar{d} := \max_j d_j$.

For every $\bar{d} \le d \in \mathbb{N}$ consider semidefinite programs indexed by d:

$$
\begin{aligned}
\rho_d = \inf_{\mathbf{y}^+,\mathbf{y}^-} \{ y_0^+ + y_0^- : \; & y_\alpha^+ - y_\alpha^- = b_\alpha, \quad \alpha \in \mathbb{N}_t^n \\
& \mathbf{M}_{d-d_j}(g_j\,\mathbf{y}^+), \mathbf{M}_{d-d_j}(g_j\,\mathbf{y}^-) \succeq 0, \quad j = 0, \ldots, m \}.
\end{aligned}
\tag{5.7}
$$

(5.7) provides a hierarchy of semidefinite programs whose associated sequence of optimal values is monotone non-decreasing. For each $d \ge \bar{d}$, the dual of (5.7) is the semidefinite program

$$
\begin{aligned}
\rho_d^* = \sup_{p,\sigma_j^1,\sigma_j^2} \{ \sum_{\alpha \in \mathbb{N}_t^n} p_\alpha\, b_\alpha : \\
1 - p = \sum_{j=0}^{m} \sigma_j^1\, g_j; \quad 1 + p = \sum_{j=0}^{m} \sigma_j^2\, g_j, \\
p \in \mathbb{R}[\mathbf{x}]_t; \; \sigma_j^i \in \Sigma[\mathbf{x}]_{d-d_j}, \; j = 0, \ldots, m; \; i = 1, 2 \}.
\end{aligned}
\tag{5.8}
$$

Theorem 5.4. *Let $\mathbf{X}$ be compact with nonempty interior.*

(i) For every $d \ge \bar{d}$ there is no duality gap between the semidefinite programs (5.7) and (5.8), i.e., $\rho_d = \rho_d^*$. In addition, both of them have an optimal solution.

(ii) Assume that (5.2) has a unique optimal solution $\hat{\phi} \in \mathcal{M}(\mathbf{X})$ with Jordan decomposition $\hat{\phi}^+ - \hat{\phi}^-$, and let $(\mathbf{y}^{+d}, \mathbf{y}^{-d})$ be an arbitrary optimal solution of (5.7). Then

$$
\lim_{d \to \infty} y_\alpha^{+d} = \int_{\mathbf{X}} \mathbf{x}^\alpha \, d\hat{\phi}^+; \quad \lim_{d \to \infty} y_\alpha^{-d} = \int_{\mathbf{X}} \mathbf{x}^\alpha \, d\hat{\phi}^-, \quad \forall \alpha \in \mathbb{N}^n.
\tag{5.9}
$$

(iii) Let $r = \max[d_j]$. If

$$
\operatorname{rank} \mathbf{M}_{d-r}(\mathbf{y}^{+d}) = \operatorname{rank} \mathbf{M}_d(\mathbf{y}^{+d})
\tag{5.10}
$$

$$
\operatorname{rank} \mathbf{M}_{d-r}(\mathbf{y}^{-d}) = \operatorname{rank} \mathbf{M}_d(\mathbf{y}^{-d})
$$

then finite convergence takes place, i.e., for every $\alpha \in \mathbb{N}^n$,

$$
y_\alpha^{+d} = \int_{\mathbf{X}} \mathbf{x}^\alpha \, d\hat{\phi}^+; \quad y_\alpha^{+-} = \int_{\mathbf{X}} \mathbf{x}^\alpha \, d\hat{\phi}^-.
\tag{5.11}
$$

Moreover, one may recover the rank $\mathbf{M}_d(\mathbf{y}^{+d})$ *atoms and weights of* $\hat{\phi}^+$ *and the* rank $\mathbf{M}_d(\mathbf{y}^{-d})$ *atoms and weights of* $\hat{\phi}^-$ *by a linear algebra routine; see Theorem 1.6.*

Proof. Brief sketch of the proof. (i) The feasible set of (5.3) is compact. This is because $g_1(\mathbf{x}) = -\|\mathbf{x}\|^2$ and therefore the sdp constraints $\mathbf{M}_d(\mathbf{y}^{\pm d}), \mathbf{M}_d(\mathbf{y}^{\pm d} g_j) \succeq 0$ imply $|\mathbf{y}_\alpha^{\pm d}| < 1$ for all $\alpha \in \mathbb{N}_{2d}^n$. Hence the set of optimal solutions of (5.3) is compact. By a result of Trnovská (2005) this also implies that there is no duality gap between (5.3) and (5.4). It remains to prove that (5.4) has an optimal solution. Let (p_n, σ_{nj}^i) be a maximizing sequence of (5.4). In the proof of Theorem 5.3 we have seen that the set U in (5.5) is compact. Therefore there is $p^* \in \mathbb{R}[\mathbf{x}]_t$ and a subsequence n_k such that $p_{n_k} \to p^*$ as $k \to \infty$. Recall that by Lemma 1.2 $Q(g)$ is closed and so as $1 \pm p_n \in Q_d(g)$ it follows that $1 \pm p^* \in Q_d(g)$.

(ii) As for each d one has $|y_\alpha^{\pm d}| \leq 1$ for all $\alpha \in \mathbb{N}_{2d}^n$, by a standard argument already used in other chapters, there is a subsequence $(d_k)_{k \in \mathbb{N}}$ and two infinite sequences $\mathbf{y}^\pm$ such that for every $\alpha \in \mathbb{N}^n$, $\lim_{k \to \infty} y_\alpha^{\pm d_k} = y_\alpha^\pm$. In addition, this convergence also ensures that $\mathbf{M}_d(g_j \, \mathbf{y}^\pm) \succeq 0$ for all d. By Putinar's Theorem 1.9 $\mathbf{y}^+$ and $\mathbf{y}^-$ have a representing measure on $\mathbf{X}$, say ϕ^+ and ϕ^-. Again by the above convergence, $y_\alpha^+ - y_\alpha^- = b_\alpha$ for all $\alpha \in \mathbb{N}_t^n$, and in addition:

$$y_0^+ + y_0^- = \lim_{k \to \infty} (y_0^{+d_k} + y_0^{-d_k}) = \lim_{k \to \infty} \rho_{d_k} \leq \rho.$$

That is, (ϕ^+, ϕ^-) is an optimal solution of (5.3), hence of (5.2). As the latter is unique and the converging subsequence $(d_k)_{k \in \mathbb{N}}$ was arbitrary it follows that the whole sequences $(\mathbf{y}^{+d})$ and $(\mathbf{y}^{-d})$ converge, which yields (5.9). Finally (iii) follows from Theorem 1.6. □

5.2.2 The Moment-SOS hierarchy in action

In practice, Theorem 5.4 should be used as follows:

1 Let $k := \bar{d}$.

2 Solve SDP problem (5.7) and its dual (5.8) with a primal-dual algorithm.

3 If the rank condition (5.10) is satisfied, then extract the measure from the solution of (5.7) and the polynomial certificate from the solution of (5.8). Otherwise, let $k = k + 1$, and go to 1.

We conjecture that if the data of the problem are generic, then there is a finite value of k for which the rank-condition of Theorem 5.4 is satisfied. The rationale behind this assertion follows from a result in Nie (2014) on *generic* finite convergence for the Moment-SOS hierarchy for polynomial optimization over compact basic semi-algebraic sets. Similarly, from Nie (2013) the rank-condition of Theorem 5.4 also holds generically for polynomial optimization (which however is a context different from the present context). For a discussion on this issues the reader is referred to de Castro et al. (2017).

5.3 Examples

In this section we present some examples for which the Moment-SOS hierarchy is the first to give an SDP formulation of the Super Resolution phenomena. As a matter of fact, our procedure encompasses a larger class of measurements than the class of standard moments discussed previously. Our numerical experiments are carried out with the Matlab interface GloptiPoly 3, a software dedicated to solving the GMP with algebraic data, which is designed to generate semidefinite relaxations of measure LP problems with polynomial data. In all examples we use monomial moments, a choice motivated only for notational simplicity. In fact other choices of polynomials (e.g. Chebyshev polynomials) are typically preferable numerically.

Example 5.1. (Disconnected domain) We want to recover the measure:

$$\phi := \delta_{-3/4} + \delta_{1/2} - \delta_{1/8}$$

on the disconnected set $\mathbf{X} := [-1, -1/2] \cup [0, 1]$ which can be modeled as the polynomial superlevel set $\mathbf{X} = \{x \in \mathbb{R} : g_1(x) \geq 0\}$ for the choice:

$$g_1(x) := -(x + 1)(x + 1/2)x(x - 1).$$

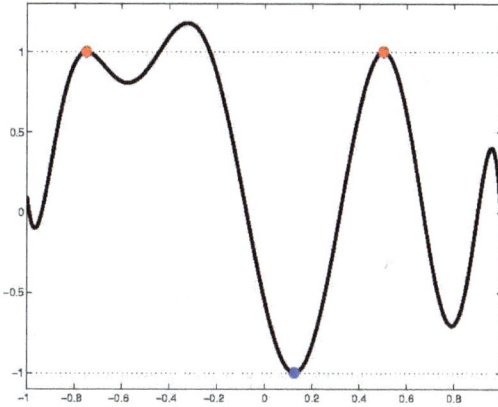

Fig. 5.1   Example 5.1: Degree-9 polynomial certificate with 2 points (red) in the support of ϕ^+ and 1 point (blue) in the support of ϕ^-.

We assume knowledge of 10 moments $b_i := (-3/4)^i + (1/2)^i - (1/8)^i$ for $i = 0, 1, 2\ldots, 9$. Solving the SDP relaxation of order $k = 5$ on a standard PC takes 0.2 seconds, and optimality is certified from the solution of the primal moment problem with a rank-2 moment matrix for ϕ^+ and a rank-1 moment matrix for ϕ^-, from which the 3 points can be extracted.

In Figure 5.1 is displayed the degree-9 polynomial certifying optimality, constructed from the solution (5.4). Indeed we can check that the polynomial attains the value $+1$ at the points $x = -3/4$, and $x = 1/2$ (in red), it attains the value -1 at the point $x = 1/4$ (in blue), while taking values between -1 and $+1$ on $\mathbf{X}$. Notice in particular that the polynomial is larger than $+1$ around $x = -1/4$, but this point is not in $\mathbf{X}$.

Example 5.2. (Low-pass filters in dimension greater than 3) In the Fourier frame, the recent SDP formulations of ℓ_1-minimization in the space of complex valued measures are based on the Fejér-Riesz theorem. As a consequence, they cannot handle dimensions greater than 3. Observe that the Moment-SOS Hierarchy can bypass this limitation. We want to recover the measure

$$\phi := \delta_{(-1/2,1/2)} + \delta_{(1/2,-1/2)} + \delta_{(1/2,1/2)} + \delta_{(0,0)} - \delta_{(0,-1/2)} - \delta_{(1/2,0)}$$

on the box $\mathbf{X} := [-1, 1]^2$ from knowledge of monomial moments up to degree 12. Solving the SDP relaxation (5.3) with $d = 6$ on a standard PC takes less than 3 seconds, and optimality is certified via (5.10) with a rank-4 moment matrix for ϕ^+ for and a rank-2 moment matrix for ϕ^- from which the 6 points of the support of the optimal measure ϕ can be extracted using numerical linear algebra with a relative accuracy around 10^{-6}. In Figure 5.2 is displayed the degree-12 polynomial certifying optimality, constructed from the solution of the dual (5.4) Indeed we can check that the polynomial attains the value $+1$ at the 3 points $x \in \{(-1/2, 1/2), (1/2, -1/2), (0, 0)\}$, it attains the value -1 at the 2 points $x \in \{(0, -1/2), (1/2, 0)\}$ (in blue), while taking values between -1 and $+1$ on $\mathbf{X}$.

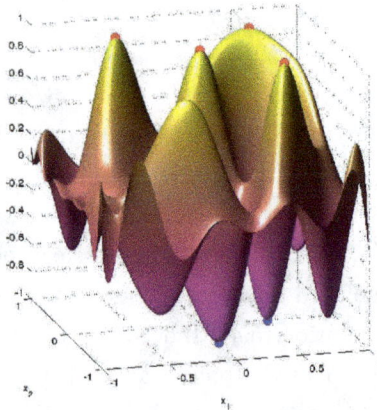

Fig. 5.2 Example 5.2: Degree-12 polynomial certificate with 4 points (red) in the support of ϕ^+ and 2 points (blue) in the support of ϕ^-.

Example 5.3. (Localization of points on the sphere) Recent extensions of Super Resolution to spike deconvolution on the 2-sphere from spherical harmonic measurements has been investigated in Bendory et al. (2014a) and Bendory et al. (2014b). In these paper, the authors give a sufficient condition for exact recovery using ℓ_1-minimization and they investigate spikes localization when the measurements are perturbed by additive noise. From a numerical point of view, they used a relaxed version of the dual program (bounded real lemma in dimension $d = 3$ and a Gram representation of the ℓ_∞-constraint appearing in the dual). Our work naturally extends to this frame and provides an exact formulation of the primal form.

For sake of numerical code simplicity, we have considered polynomials on the 2-sphere $\mathbf{X} \subset \mathbb{R}^3$ (note that one could have used homogenous spherical harmonics instead as in Bendory et al. (2014a)). We want to recover the measure

$$\phi := \delta_{(1,0,0)} + \delta_{(0,1,0)} + \delta_{(0,0,1)} - \delta_{(\frac{\sqrt{2}}{2}, \frac{\sqrt{2}}{2}, 0)} - \delta_{(\frac{\sqrt{2}}{2}, 0, \frac{\sqrt{2}}{2})} - \delta_{(0, \frac{\sqrt{2}}{2}, \frac{\sqrt{2}}{2})}$$

supported on the positive orthant just for better visualization purposes. All 56 monomial moments up to degree 5 are assumed to be available. Solving the SDP relaxation of order $d = 6$ on a standard PC takes less than 20 seconds, and optimality is certified from the solution of the primal moment problem with a rank-3 moment matrix for ϕ^+ and a rank-3 moment matrix for ϕ^-, from which the 6 points can be extracted; see Theorem 1.6. In Figure 5.3 is displayed the degree-6 polynomial that certifies optimality on the 2-sphere, constructed from the solution of the dual (5.4). Indeed one can check that the polynomial attains the value $+1$ at the 3 prescribed points, attains the value -1 (in red), at the 3 others prescribed points (in blue), while taking values between -1 and $+1$ on $\mathbf{X}$. From a theoretical point of view, the minimal separation condition appearing in Bendory et al. (2014a) requires a degree-2 polynomial. So our example satisfy the sufficient aforementioned condition.

5.4 Concluding remarks

In this chapter we have shown that retrieving an atomic measure (with finite support on a compact basic semi-algebraic set) can be reconstructed from the sole knowledge of a few of its moments via the Moment-SOS hierarchy. It extends to a multivariate setting the seminal work of Candès and Fernandez-Granda (2014) on the torus (although the multivariate case is briefly mentioned there). On the torus, finite convergence of the Moment-SOS hierarchy follows from the Fejér-Riesz theorem for

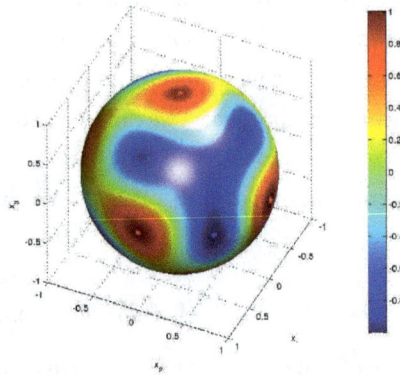

Fig. 5.3   Example 5.3: Degree-6 polynomial certificate with 3 points (red) in the support of ϕ^+ and 3 points (blue) in the support of ϕ^-.

trigonometric polynomials on $\mathbf{X} = [0, 2\pi]$, and this was exploited in the landmark paper of Candès and Fernandez-Granda (2014). It turns out that the Fejér-Riesz theorem also holds in dimension $n = 2$ which also guarantees finite convergence of the Moment-SOS on the bivariate torus (but now the step at which finite convergence takes place is not known in advance). This result (apparently not so well-known) follows from Corollary 3.4 in Scheiderer (2006).

5.5   Notes and sources

This chapter is essentially from de Castro et al. (2017). As already mentioned the early formulation of the Super Resolution problem can be identified as the ability of faithfully reconstruct a high-dimensional sparse vector from the observation of a low-pass filter. This situation models important applications in imaging spectroscopy Harris et al. (1994), image processing Park et al. (2003), radar imaging Odendaal et al. (1994), or astronomy Makovoz and Marleau (2005). To the best of our knowledge, the first mathematical works on Super Resolution are due to Donoho and his collaborators in the early nineties; see e.g., Donoho and Stark (1989) and Donoho (1992). The problem is posed as retrieving a sparse vector $\mathbf{x}^*$, solution of an under-determined linear system $\mathbf{A}\mathbf{x} = \mathbf{b}\,(= \mathbf{A}\mathbf{x}^*)$ and the term *Super Resolution* appeared because the matrix $\mathbf{A}$ is related to a discretization of some Fourier transform. As a matter of fact, when inverting a discrete Fourier transform, the separation of two close spikes in a sparse signal is made possible by minimizing the ℓ_1-norm while a linear inversion method is not able to do so. These works have seeded the ideas of compressed sensing theory by Donoho (2006a), Candès and Tao (2006), and Candès and Tao (2007). In this theory, the matrix $\mathbf{A}$ is randomized and one is interested both in the construction of probability distributions allowing to show relevant properties, such as the Restricted Isometry Property of Candès and Tao (2006), and the stability of the reconstruction process. The seminal paper of Candès and Fernandez-Granda (2014) showed that an atomic (signed) measure with finite support on the torus is the unique optimal solution of an infinite-dimensional LP provided that its atoms are sufficiently separated and enough moments are available.

Finally, it is worth pointing out that the work in de Castro et al. (2017) was also inspired by a previous work on optimal control for linear systems formulated as the primal LP on measures (5.2) and a dual LP on continuous functions like (5.4), and solved numerically with primal-dual moment-SOS SDP hierarchies in Claeys et al. (2014). Formulating optimal control problems as moment problems was a classical research topic in the 1960s, where optimal control laws were sought in measures spaces (completions of Lebesgue spaces) to allow for oscillations and concentrations, see e.g. Krasovskii (1968) or the overview in (Fattorini, 1999, Section III). In the case of linear optimal control of an ordinary differential equation of order n, it was proved in Neustadt (1964) that there is always an n-atomic optimal measure solving problem (5.2).

Bibliography

Azais, J.-M., Y. de Castro, Y., Gamboa, F. (2014). Spike detection from inaccurate samplings, Appl. Comput. Harmonic Anal.

Barvinok, A. (2002). *A course in convexity*, Graduate Studies in Mathematics, vol. 54, American Mathematical Society, Providence, RI.

Bendory, T., Dekel, S., Feuer, A. (2014a). *Exact recovery of Dirac ensembles from the projection onto spaces of spherical harmonics*, Const. Approx., pp. 1–25.

Bendory, T., Dekel, S., Feuer, A. (2014a). *Super resolution on the sphere using convex optimization*, Preprint, arXiv:1412.3282.

Claeys, M., Arzelier, D., Henrion, D., Lasserre, J. B. (2014). *Moment LMI approach to LTV impulsive control*, IEEE Trans. Aut. Control **59**, pp. 1374–1379.

Candès E. J., and Fernandez-Granda, C. (2014). *Towards a mathematical theory of superresolution*, Comm. Pure Appl. Math. **67**, pp. 906–956.

Candès, E. J., and Tao, T. (2006). *Near-optimal signal recovery from random projections: universal coding strategies?*, IEEE Trans. Info. Theory **52**, pp. 5406–5425.

Candès, E. J., and Tao, T. (2007). *The Dantzig selector: statistical estimation when p is much larger than n*, Ann. Statist. **35**, pp. 2313–2351.

de Castro, Y., Gamboa, F., Henrion D., Lasserre, J. B. (2017). *Exact solutions to Super Resolution on semi-algebraic domains in higher dimensions*, IEEE Trans. Info. Theory **63**, pp. 621–630.

de Castro, Y., and Gamboa, F. (2012). *Exact reconstruction using Beurling minimal extrapolation*, J. Math. Anal. Appl. **395**, pp. 336–354.

Doukhan, P., and Gamboa, F. (1996). *Superresolution rates in Prokhorov metric*, Canad. J. Math. **48**, pp. 316–329.

Donoho, D. L. (1992). *Superresolution via sparsity constraints*, SIAM J. Math. Anal. **23**, pp. 1309–1331.

Donoho, D. L. (2006a). *Compressed sensing*, IEEE Trans. Info. Theory **52**, pp. 1289–1306.

Donoho, D. L. (2006b). *For most large underdetermined systems of linear equations the minimal ℓ_1-norm solution is also the sparsest solution*, Comm. Pure Appl. Math. **59**, pp. 797–829.

Duval, V., and Peyré, G. (2013). *Exact support recovery for sparse spikes deconvolution*, Found. Comput. Math. **7**, pp. 1–41.

Donoho, D. L., and Stark, P. B. (1989). *Uncertainty principles and signal recovery*, SIAM J. Appl. Math. **49**, pp. 906–931.

Fattorini, H. O. (1999). *Infinite dimensional optimization and control theory* (Cambridge University Press, UK).

Harris, T. D., Grober, R. D., Trautman, J. K., Betzig, E. (1994). *Super-resolution imaging, spectroscopy*, Appl. Spectroscopy **48**(1), pp. 14A–21A.

Josz, C., Lasserre J. B., Mourrain, B. (2019). *Sparse polynomial interpolation: Sparse recovery, super-resolution or Prony?*, Adv. Comp. Math. **45**, pp. 1401–1437.

Krasovskii, N. N. (1968). *Theory of motion control: Linear Systems*, Nauka, Moscow.

Lasserre, J. B. (2010). *Moments, Positive Polynomials and Their Applications*, Imperial College Press Optimization Series, vol. 1, Imperial College Press, London, UK.

Makovoz D., and Arleau, F. R. (2005). *Point-source extraction with mopex*, Publications of the Astronomical Society of the Pacific **117**, pp. 1113–1128.

Neustadt, L. W. (1964). *Optimization, a moment problem, and nonlinear programming*, J. Soc. Indus. Appl. Math. Series A: Control **2**, pp. 33–53.

Nie, J. (2013). *Certifying convergence of Lasserre's hierarchy via flat truncation*, Math. Program. Ser. A **142**, pp. 485–510.

Nie, J. (2014). *Optimality conditions and finite convergence of Lasserre's hierarchy*, Math. Program. Ser. A **146**, pp. 97–121.

Odendaal, J. W., Barnard, E., Pistorius, C. W. I. (1994). *Two-dimensional super-resolution radar imaging using the music algorithm*, IEEE Trans. Antennas Propagation **42**, pp. 1386–1391.

Park, S. C., Park, M. K., Kang, M. G. (2003). *Super-resolution image reconstruction: A technical overview*, IEEE Trans. Signal Proc. Magazine, pp. 21–36.

Scheiderer, C. (2006). *Sums of squares on real algebraic surfaces*, Manuscipta Mathematica **119**, pp. 395–410.

Trnovská, M. (2005). *Strong duality conditions in semidefinite programming*, J. Elec. Eng. **56**, pp. 1–5.

Chapter 6

Sparse Polynomial Interpolation

We describe how to apply the Moment-SOS hierarchy to retrieve the coefficients of a "black-box" polynomial known only through a few evaluations.

6.1 Introduction

Suppose that we are given a *black-box* polynomial $g \in \mathbb{R}[\mathbf{z}]$, that is, g is unknown but given any "input" point $\mathbf{z} \in \mathbb{C}^n$, the black-box outputs the complex number $g(\mathbf{z})$. A *sparse* polynomial is a polynomial $\mathbf{z} \mapsto g(\mathbf{z}) = \sum_\alpha g_\alpha \mathbf{z}^\alpha$ with only a few non-zero coefficients $\mathbf{g} = (g_\alpha)$.

> *Sparse interpolation* is concerned with recovering the unknown monomials $(\mathbf{z}^\alpha)$ and coefficients (g_α) of a sparse polynomial via the sole knowledge of a few (and as few as possible) values of g at some points $(\mathbf{z}_k) \subset \mathbb{C}^n$ that one may choose at our convenience.

Basically there are two possible methods to recover the coefficients of the unknown polynomial:

Naive LP. A naive "compressed sensing" LP-approach consists of solving $\min\{\|\mathbf{x}\|_1 : \mathbf{A}\mathbf{x} = \mathbf{b}\}$ where $\mathbf{x}$ is the vector of coefficients of the unknown polynomial and $\mathbf{A}\mathbf{x} = b$ are linear constraints obtained from evaluations at some given points. In minimizing the ℓ_1 norm one expects to obtain a "sparse" solution to the undetermined system $\mathbf{A}\mathbf{x} = \mathbf{b}$ (as in compressed sensing). This approach was briefly mentioned in Chapter 5. However since the matrix $\mathbf{A}$ does not satisfy the sufficient *Restricted Isometry Property* (RIP), exact recovery is not guaranteed (at least by invoking results from compressed sensing).

Prony. The method goes back to the pioneer work of G. R de Prony Prony (1795) who was interested in recovering a sum of few exponential terms from sampled values of the function. Thus Prony's method is also a standard tool to recover a complex atomic measure from knowledge of some of its moments; see e.g. Kunis et al. (2016). Briefly, in the univariate setting this purely algebraic method consists of two steps: (i) Computing the coefficients of a polynomial p whose roots form the finite support of the unknown measure. As p satisfies a recurrence relation it is the unique element (up to scaling) in the kernel of a (Hankel) matrix. (ii) The weights associated to the atoms of the support solve a Vandermonde system.

This algebraic method has then been used in the context of sparse polynomial interpolation. In the univariate case it consists in evaluating the black-box polynomial at values of the form φ^k for a finite number of pairs $(k, \varphi) \in \mathbb{N} \times \mathbb{C}$, fixed. A sequence of $2r$ evaluations allows to recover the decomposition exactly, where r is the number of terms of the sparse polynomial. The decomposition is obtained by computing a minimal recurrence relation between these evaluations, by finding the roots of the associate polynomial, which yields the exponents of the monomials and by solving a Vandermonde system which yields the coefficients of the terms in the sparse polynomial. Then Prony has been generalized to multivariate reconstruction problems.

6.2 Sparse interpolation as a super-resolution problem

We show that sparse interpolation is in fact a Super Resolution problem on the torus $\mathbb{T}^n \subset \mathbb{C}^n$, provided that evaluations are made at points chosen in a certain adequate manner. So let

$$\mathbf{x} \mapsto g(\mathbf{z}) = \sum_{\beta \in \mathbb{N}_d^n} g_\beta \, \mathbf{z}^\beta,$$

be the black-box polynomial with unknown real coefficients $(g_\beta) \subset \mathbb{R}$.

A crucial observation. Let $\mathbf{z}_0 \in \mathbb{T}^n$ (with $\mathbf{z}_0 \neq (1, \ldots, 1)$) be fixed, e.g., of the form:

$$\mathbf{z}_0 := (\exp(2i\pi/N), \ldots, \exp(2i\pi/N)), \tag{6.1}$$

for some arbitrary (fixed) integer N, or

$$\mathbf{z}_0 := (\exp(2i\pi\theta_0), \ldots, \exp(2i\pi\theta_0)), \tag{6.2}$$

for some arbitrary (fixed) irrational $\theta_0 \in \mathbb{R}$. With the choice (6.2)

$$[\, \alpha, \beta \in \mathbb{Z}^n \text{ and } \alpha \neq \beta \,] \quad \Rightarrow \quad \mathbf{z}_0^\alpha \neq \mathbf{z}_0^\beta$$

whereas with the choice (6.1)

$$[\, \alpha, \beta \in \mathbb{Z}^n, \max_i \max[|\alpha_i|, |\beta_i|] < N, \text{ and } \alpha \neq \beta \,] \quad \Rightarrow \quad \mathbf{z}_0^\alpha \neq \mathbf{z}_0^\beta.$$

Next for every $\alpha \in \mathbb{N}^n$:

$$g(\mathbf{z}_0^\alpha) = \sum_{\beta \in \mathbb{N}^n} g_\beta (\mathbf{z}_0^\alpha)^\beta = \sum_{\beta \in \mathbb{N}^n} g_\beta (z_{01}^{\alpha_1})^{\beta_1} \cdots (z_{0n}^{\alpha_n})^{\beta_n} \tag{6.3}$$

$$= \sum_{\beta \in \mathbb{N}^n} g_\beta (z_{01}^{\beta_1})^{\alpha_1} \cdots (z_{0n}^{\beta_n})^{\alpha_n}$$

$$= \sum_{\beta \in \mathbb{N}^n} g_\beta (\mathbf{z}_0^\beta)^\alpha = \int_{\mathbb{T}^n} \mathbf{z}^\alpha \, d\mu_{g, \mathbf{z}_0}(\mathbf{z}),$$

where $\mu_{g, \mathbf{z}_0}$ is the signed atomic-measure on $\mathbb{T}^n$ defined by:

$$\mu_{g, \mathbf{z}_0} := \sum_{\beta \in \mathbb{N}^n} g_\beta \, \delta_{\xi_\beta} \quad (\text{and } \|\mu_{g, \mathbf{z}_0}\|_{TV} = \sum_\beta |g_\beta| = \|g\|_1), \tag{6.4}$$

where $\xi_\beta := \mathbf{z}_0^\beta \in \mathbb{T}^n$, for all $\beta \in \mathbb{N}^n$ such that $g_\beta \neq 0$, and δ_ξ is the Dirac probability measure at the point $\xi \in \mathbb{C}^n$. Note that in (6.3) we slightly abuse by using the same notation $\mathbf{z}_0^\alpha$ for both the point $(z_{01}^{\alpha_1}, \ldots, z_{0n}^{\alpha_n}) \in \mathbb{C}^n$ and the monomial $z_{01}^{\alpha_1} \cdots z_{0n}^{\alpha_n}$; the context makes clear which one is correct.

In other words: *Evaluating g at the point $\mathbf{z}_0^\alpha \in \mathbb{T}^n$ is the same as evaluating the moment $\int_{\mathbb{T}^n} \mathbf{z}^\alpha \, d\mu_{g,\mathbf{z}_0}$ of the signed atomic-measure $\mu_{g,\mathbf{z}_0}$. Therefore, the sparse interpolation problem is the same as recovering the finitely many unknown weights $(g_\beta) \subset \mathbb{R}$ and supports $(\mathbf{z}_0^\beta) \subset \mathbb{T}^n$ of the signed measure $\mu_{g,\mathbf{z}_0}$ on $\mathbb{T}^n$, from finitely many s moments of $\mu_{g,\mathbf{z}_0}$, that is, a Super Resolution problem as described in Chapter 5.*

Remark 6.1. The n-dimensional torus $\mathbb{T}^n$ is one among possible choices but any other choice of a set $S \subset \mathbb{C}^n$ and $\mathbf{z}_0 \in \mathbb{C}^n$ (or $S \subset \mathbb{R}^n$ and $\mathbf{z}_0 \in \mathbb{R}^n$) is valid provided that $(\mathbf{z}_0^\alpha)_{\alpha \in \mathbb{N}^n} \subset S$. For instance $\mathbf{z}_0 \in (-1,1)^n$ and $S := [-1,1]^n$ is another possible choice.

6.2.1 A uniqueness result

Given $\mathfrak{d} \in \mathbb{N}$, define the set

$$\mathscr{A}_\mathfrak{d}^1 = \{\alpha \in \mathbb{Z}^n \mid \|\alpha\|_1 = \sum_i |\alpha_i| \leq \mathfrak{d}\}, \qquad (6.5)$$

and let $\mathbb{C}[\mathbf{z}; \mathscr{A}_\mathfrak{d}^1] := \{\sum_{\alpha \in \mathscr{A}_\mathfrak{d}^1} \theta_\alpha \mathbf{z}^\alpha : \theta \in \mathbb{C}\}$.

With the choice $S := \mathbb{T}^n$, $\mathbf{z}_0 \in \mathbb{T}^n$ as in (6.1) or in (6.2), $\mathfrak{d} \in \mathbb{N}$ and $\mathscr{A}_\mathfrak{d}^1$ in (6.5), consider the optimization problem:

$$\rho_\mathfrak{d} = \inf_{\phi \in \mathscr{M}(\mathbb{T}^n)} \left\{ \|\phi\|_{TV} : \int_{\mathbb{T}^n} \mathbf{z}^\alpha \, d\phi(\mathbf{z}) = a_\alpha, \quad \alpha \in \mathscr{A}_\mathfrak{d}^1 \right\}, \qquad (6.6)$$

where $a_\alpha = g(\mathbf{z}_0^\alpha)$ is obtained from the black-box polynomial $g \in \mathbb{R}[\mathbf{z}]$, and

$$\rho_\mathfrak{d}^* = \sup_{g \in \mathbb{C}[\mathbf{z}; \mathscr{A}_{f_c}]} \left\{ \Re(\mathbf{a}^T \mathbf{g}) : \|\Re(g(\mathbf{z}))\|_\infty \leq 1 \right\} \qquad (6.7)$$

(where $\mathbf{a} = (a_\alpha)$ and $\mathbf{g} = (g_\alpha)$).

In (6.6) we recognize the Super Resolution problem treated in (5.2) except that now the support of the unknown measure is $\mathbb{T}^n \subset \mathbb{C}^n$. By Theorem 5.2 (adapted to $\mathbb{T}^n$) (6.6) is equivalent to the infinite-dimensional

LP:

$$\rho_{\mathfrak{d}} = \inf_{\phi^+,\phi^- \in \mathcal{M}(\mathbb{T}^n)_+} \{ \int_{\mathbb{T}^n} d(\phi^+ + \phi^-) :$$
$$\text{s.t.} \int_{\mathbb{T}^n} \mathbf{z}^\alpha \, d(\phi^+ - \phi^-) = a_\alpha, \quad \alpha \in \mathscr{A}_{\mathfrak{d}}^1 \}, \tag{6.8}$$

with same dual (6.7). We next prove that the minimization problem (6.6) has a unique solution, provided that $\mathfrak{d}$ is sufficiently large.

For $d \in \mathbb{N}$, let $\mathbb{A}_d$ be the vector space spanned by the monomials $\mathbf{z}^\alpha$ with $\alpha \in \mathscr{A}_d^1$. For $\Xi = \{\xi_1, \dots, \xi_r\} \subset \mathbb{T}^n$, we denote by $\mathcal{E}(\Xi)$, the lowest $M = \max\{|u_i(\xi_r)|^2\}$ for all the families of interpolation polynomials $u_1, \dots, u_r \in \mathbb{C}[\mathbf{z}^{\pm 1}]$ of total degree $\leq r - 1$. We call $\mathcal{E}(\Xi)$ the *interpolation entropy* of Ξ. By standard arguments on the quotient algebra by an ideal defining r points, it is always possible to find a family of interpolation polynomials of total degree $\leq r - 1$. Notice that $\mathcal{E}(\Xi)$ is related to the condition number of the Vandermonde matrix of the monomial basis of $\mathbb{A}_{r-1}$ at the points Ξ. Thus it depends on the separation of these points.

To prove that the minimization problem (6.6) has a unique solution, we first prove the existence of a dual *certificate*.

Lemma 6.2. *Let $\Xi = \{\xi_1, \dots, \xi_r\} \subset \mathbb{T}^n$ and $\epsilon_i \in \{\pm 1\}$. Let $\mathfrak{d} \geq 4\,\pi\,r(r - 1)\,\mathcal{E}(\Xi)$. There exists $q(\mathbf{z}) \in \mathbb{A}_{\mathfrak{d}}$ such that:*

- *$q(\xi_i) = \epsilon_i$ for $i = 1, \dots, r$,*
- *$|q(\mathbf{z})| < 1$ for $\mathbf{z}$ in an open dense subset of $\mathbb{T}^n$.*

Proof. Let $u_i(\mathbf{z}) \in \mathbb{C}[\mathbf{z}^{\pm 1}]$, $i = 1, \dots, r$ be a family of interpolation polynomials at Ξ, with support in $\mathbb{A}_{r-1}$ and which reaches $\mathcal{E}(\Xi)$. They satisfy the following properties: $U_i(\xi_j) = \delta_{i,j}$ for $1 \leq i, j \leq r$. We denote $M = \max_{\mathbf{z} \in T^n} \{|u_i(\mathbf{z})|^2, i = 1, \dots, r\}$. For $v_1, \dots, v_r \in \mathbb{R}$, let

$$U(\mathbf{z}) = \sum_{i=1}^{r} v_i \, u_i(\mathbf{z}) \overline{u}_i(\mathbf{z}^{-1}). \tag{6.9}$$

It is a polynomial with support in $\mathbb{A}_{\mathfrak{d}}$ and with real values $\sum_{i=1}^{r} v_i |u_i(\mathbf{z})|^2$ for $\mathbf{z} \in \mathbb{T}^n$. A direct computation shows that

$$\max_{\mathbf{z} \in T^n} |U(\mathbf{z})| \leq rMm$$

where $m = \max_i\{|v_i|\}$. For $m \le \frac{1}{rM}$ and $\mathbf{z} \in \mathbb{T}^n$, $|U(\mathbf{z})| \le 1$.

Choose a Chebyshev polynomial $T(x)$ of degree d big enough so that it has 2 extremal points $\zeta_-, \zeta_+ \in]-\frac{1}{rM}, \frac{1}{rM}[$ with $T(\zeta_-) = -1$ and $T(\zeta_+) = 1$. For instance choose d such that $\frac{\pi}{d} \le \frac{1}{2rM}$, i.e., $d \ge 2\pi rM$. On $[-1,1]$ and except at the roots of $T'(x) = 0$, the norm $|T(x)|$ is strictly less than 1.

Let $U(\mathbf{z})$ be the polynomial (6.9) constructed with $v_i = \zeta_+$ if $\epsilon_i = 1$ and $v_i = \zeta_-$ if $\epsilon_i = -1$, and let $q(\mathbf{z}) = T(U(\mathbf{z})) \in \mathbb{C}[\mathbf{z}^{\pm 1}]$. As $u_i \in \mathbb{A}_{r-1}$ and $d \ge 2\pi r \mathcal{E}(\Xi)$, $q(\mathbf{z}) \in \mathbb{A}_{\mathfrak{d}}$ for $\mathfrak{d} \ge 4\pi r(r-1)\mathcal{E}(\Xi)$.

Then we have $q(\xi_i) = T(\zeta_{\epsilon_i}) = \epsilon_i$ and for $\zeta \in \mathbb{T}^n$, $|q(\mathbf{z})| \le 1$ since $|U(\mathbf{z})| \in [-1,1]$ and $|T(x)| \le 1$ for $x \in [-1,1]$. Moreover, for $\mathbf{z} \in \mathbb{T}^n$, $|q(\mathbf{z})| = 1$ only when $U(\mathbf{z})$ reaches a root of $T'(x)$ on $[-1,1]$. This cannot be the case on a dense open subset of $\mathbb{T}^n$, since $U(\mathbf{z})$ is a non-constant polynomial. Thus $|q(\mathbf{z})| < 1$ for $\mathbf{z}$ in a dense open subset of $\mathbb{T}^n$. $\qquad\square$

We can now prove the unicity of the minimizer of (6.6), by an argument similar to the one used in Candès and Fernandez-Granda (2014).

Theorem 6.3. *Let* $\rho = \sum_i \omega_i \delta_{\xi_i}$ *be a measure supported on points* $\Xi = \{\xi_1,\ldots,\xi_r\} \subset \mathbb{T}^n$ *with* $\omega_i \in \mathbb{R}$. *Let* $\mathfrak{d} \ge 2\pi(r-1)\,r\mathcal{E}(\Xi)$ *and let* $g_1,\ldots,g_m$ *be a basis of* $\mathbb{A}_{\mathfrak{d}}$ *and* $a_k = \int_{\mathbb{T}^n} g_k\,d\rho$, $k = 1,\ldots,m$. *The optimal solution of*

$$\inf_{\mu \in \mathbf{M}(\mathbb{T}^n)} \left\{ \|\mu\|_{TV} : \int_{\mathbb{T}^n} g_k\,d\mu = a_k, \quad k = 1,\ldots,m \right\} \qquad (6.10)$$

is the measure ρ.

Proof. Let ρ^* be the optimal solution of (6.10). It can be decomposed as $\rho^* = \rho + \nu$. The Lebesgue decomposition of ν w.r.t. ρ is of the form $\nu = \nu_\Xi + \nu_c$ where ν_Ξ is supported on Ξ and ν_c is supported on $\mathbb{T}^n \backslash \Xi$. By Radon-Nikodym Theorem, ν_Ξ has a density h w.r.t. ρ. Let $\epsilon_i = \mathrm{sign}(h(\xi_i)\,\omega_i) \in \{\pm 1\}$, $i = 1,\ldots,r$ and let $q(\mathbf{z}) \in \mathbb{C}[\mathbf{z}^{\pm 1}]$ be the polynomial constructed from Ξ and $\epsilon_1,\ldots,\epsilon_r$ as in Lemma 6.2. We have

$$\int_{\mathbb{T}^n} q(\mathbf{z})\,d\nu_\Xi = \sum_{i=1}^{r} \epsilon_i h(\xi_i)\,\omega_i = \sum_{i=1}^{r} |h(\xi_i)||\omega_i| = \|\nu_\Xi\|_{TV}.$$

As the moments of monomials in $\mathbb{A}_{\mathfrak{d}}$ are the same for ρ and ρ^*,

$$0 = \int_{\mathbb{T}^n} q(\mathbf{z})\,d\nu = \int_{\mathbb{T}^n} q(\mathbf{z})\,d\nu_{\Xi} + \int_{\mathbb{T}^n} q(\mathbf{z})\,d\nu_c$$

$$= ||\nu_{\Xi}||_{TV} + \int_{\mathbb{T}^n} q(\mathbf{z})\,d\nu_c. \qquad (6.11)$$

Since $|q(\mathbf{z})| < 1$ on a dense open subset of $\mathbb{T}^n$, if $\nu_c \neq 0$ then $|\int_{\mathbb{T}^n} q(\mathbf{z})d\nu_c| < ||\nu_c||_{TV}$ and $||\nu_{\Xi}||_{TV} < ||\nu_c||_{TV}$. Hence assuming $\nu_c \neq 0$,

$$||\rho||_{TV} \geq ||\rho^*||_{TV} = ||\rho + \nu||_{TV} = ||\rho + \nu_{\Xi}||_{TV} + ||\nu_c||_{TV}$$

$$\geq ||\rho||_{TV} - ||\nu_{\Xi}||_{TV} + ||\nu_c||_{TV} > ||\rho||_{TV},$$

which yields a contradiction. Thus $\nu_c = 0$, which by (6.11) implies $||\nu_{\Xi}||_{TV} = 0$, which in turn yields $\nu = 0$ and $\rho^* = \rho$. □

The following result, a consequence of Theorem 6.3, extends to the multi-torus $\mathbb{T}^n$, $n \in \mathbb{N}$, a result of Candès and Fernandez-Granda (2014) for Super Resolution on $\mathbb{T}$ or $\mathbb{T}^2$. It is here translated in the present context of polynomial interpolation. In particular we use the set $\mathscr{A}_{\mathfrak{d}}^1$ which is smaller than $\mathscr{A}_{\mathfrak{d}} = \{\alpha \in \mathbb{Z}^n : ||\alpha||_\infty \leq \mathfrak{d}\}$ used in Candès and Fernandez-Granda (2014) for $n = 1, 2$.

Theorem 6.4. *Let $g^* \in \mathbb{R}[\mathbf{z}]$, $\mathbf{x} \mapsto g^*(\mathbf{z}) := \sum_\beta g_\beta^* \mathbf{z}^\beta$, be an unknown real polynomial. Let $\Gamma := \{\beta \in \mathbb{N}^n : g_\beta^* \neq 0\}$ and $s := |\Gamma|$. Let $\mathbf{z}_0 \in \mathbb{T}^n$ be as in (6.2) or in (6.1) (in which case $N > \max_{i=1,\ldots,n} \max\{\beta_i : \beta \in \Gamma\}$), and let $a_\alpha = g(\mathbf{z}_0^\alpha)$, $\alpha \in \mathscr{A}_{\mathfrak{d}}^1$.*

Let $\delta = \mathcal{E}(\varphi^\Gamma) = \mathcal{E}(\{\varphi^\alpha \mid \alpha \in \Gamma\})$. There is a constant $C_n > 0$ (that depends only on the dimension[1] n) such that if $\delta > C_n/\mathfrak{d}$ then the optimization problem (6.6) has a unique optimal solution ϕ^* such that

$$\phi^* := \sum_{\alpha \in \Gamma} g_\beta^* \delta_{\mathbf{z}_0^\beta} \quad and \quad ||\phi^*||_{TV} = \sum_\alpha |g_\beta^*|. \qquad (6.12)$$

In addition, there is no duality gap (i.e., $\rho_{\mathfrak{d}} = \rho_{\mathfrak{d}}^*$), (6.7) has an optimal

---

[1]If $\mathfrak{d} \geq 128$ then $C_1 = 1.87$, $C_2 = 2.38$ as proved in Candès and Fernandez-Granda Candès and Fernandez-Granda (2014).

solution $f^* \in \mathbb{C}[\mathbf{x}; \mathscr{A}_{\mathfrak{d}}^1]$, and

$$g_\beta^{*+} > 0 \Rightarrow \Re(f^*(\mathbf{z}_0^\beta)) = 1; \quad g_\beta^{*-} > 0 \Rightarrow \Re(f^*(\mathbf{z}_0^\beta)) = -1. \qquad (6.13)$$

6.3 The Moment-SOS hierarchy

Recall that in the Super-Resolution model described in Candès and Fernandez-Granda Candès and Fernandez-Granda (2014), one has to make evaluations in the multivariate case at all points $\mathbf{z}_0^\alpha$ with $\alpha \in \mathscr{A}_{\mathfrak{d}} := \{\alpha : \|\alpha\|_\infty \leqslant \mathfrak{d}\} \subset \mathbb{Z}^n$. This makes perfect sense in such applications as image reconstruction from measurements (typically 2-dimensional objects) of signal processing. However, for polynomial interpolation the size of $\mathscr{A}_{\mathfrak{d}}^\infty|$ is rapidly prohibitive if one consider polynomials of say $n = 10$ variables. Indeed, if $n = 10$ then the first-order semidefinite program of the hierarchy entails matrix variables of size 1024×1024. Bear in mind that currently, semidefinite programming solvers are limited to matrices of size a few hundred. Thus it is not possible to compute *even* the first-order relaxation!

One can reduce the computational burden by using the ℓ_1-norm truncation, i.e., $\|\alpha\|_1 := \sum_i |\alpha_i|$, by making evaluations at all points $\mathbf{z}_0^\alpha$ with

$$\alpha \in \mathscr{A}_{\mathfrak{d}}' := \{\alpha - \beta : \alpha, \beta \in \mathbb{N}^n, \|\alpha\|_1, \|\beta\|_1 \leqslant \mathfrak{d}\}. \qquad (6.14)$$

An illustration is provided in Figure 6.1. In addition, $\mathscr{A}_{\mathfrak{d}}^1 \subset \mathscr{A}_{\mathfrak{d}}'$ and:

$$\forall \mathfrak{d} \in \mathbb{N}, \quad \exists \tilde{\mathfrak{d}} \in \mathbb{N}: \quad \forall l \geqslant \tilde{\mathfrak{d}}, \quad \mathscr{A}_{\mathfrak{d}} \subset \mathscr{A}_l'. \qquad (6.15)$$

Theorem 6.4 is still valid with $\mathscr{A}_{\mathfrak{d}}'$ as in (6.14) in lieu of $\mathscr{A}_{\mathfrak{d}}^1$ (because $\mathscr{A}_{\mathfrak{d}}^1 \subset \mathscr{A}_{\mathfrak{d}}'$). To appreciate the gain in using $\mathscr{A}_{\mathfrak{d}}'$ instead of $\mathscr{A}_{\mathfrak{d}}$, for 10 variables, the first-order semidefinite program of the hierarchy entails matrix variables of size 11×11 (instead of 1024×1024) and 56 linear equalities. It is thus possible to compute the first-order relaxation. The second-order relaxation entails matrices of size 66×66 (instead of 59049×59049) and 1596 linear equalities.

A hierarchy of semidefinite programs. For every fixed $\mathfrak{d}$, and with $d \geq \mathfrak{d}$, consider the semidefinite program:

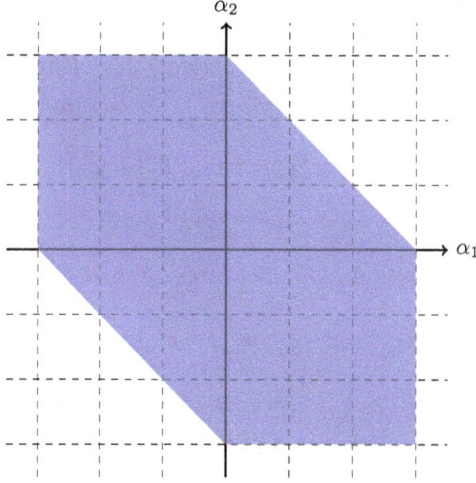

Fig. 6.1   Evaluations at $\alpha - \beta$ with $|\alpha_1| + |\alpha_2| \leqslant 3$ and $|\beta_1| + |\beta_2| \leqslant 3$ and $\alpha_1, \alpha_2, \beta_1, \beta_2 \in \mathbb{Z}$.

$$
\begin{aligned}
\rho_d = \inf_{\mathbf{y}^+, \mathbf{y}^-} \; \{\, y_0^+ + y_0^- \\
(\mathbf{P}_{d,\eth}) \qquad \text{s.t. } y_\alpha^+ - y_\alpha^- = a_\alpha \,, \quad \forall \alpha \in \mathscr{A}_\eth' \\
\mathbf{T}_d(\mathbf{y}^+) \succeq 0, \; \mathbf{T}_d(\mathbf{y}^-) \succeq 0 \,\},
\end{aligned}
\qquad (6.16)
$$

where the Hermitian matrix $\mathbf{T}_d(\mathbf{y}^+)$ has its rows and columns indexed in $\{\alpha \in \mathbb{N}^n : \|\alpha\|_1 \leqslant d\}$ and $\mathbf{T}_d(\mathbf{y}^+)_{\alpha,\beta} = y_{\beta-\alpha}^+$, for every $\|\alpha\|_1, \|\beta\|_1 \leqslant d$, and similarly for the Hermitian matrix $\mathbf{T}_d(\mathbf{y}^-)$.

In the univariate case $\mathbf{T}_d(\mathbf{y}^+)$ and $\mathbf{T}_d(\mathbf{y}^-)$ are *Toeplitz* matrices. When $\mathbf{y}^+$ is coming from a measure ϕ^+ on $\mathbb{T}^n$ then $y_\alpha^+ = \int_{\mathbb{T}^n} \mathbf{z}^\alpha \, d\phi^+(\mathbf{z})$. Clearly, (6.16) is a relaxation of (6.8) (with $\mathscr{A}_\eth'$ as in (6.14) in lieu of $\mathscr{A}_\eth^1$) and so $\rho_d \leq \rho_\eth$ for all $d \geq \eth$. Moreover $\rho_d \leq \rho_{d+1}$ for all d.

Note that with the above notations, the "Toeplitz Prony" method proposed in Josz et al. (2019) consists in directly extracting a measure from the matrix $\mathbf{T}_d(\mathbf{a})$. In contrast, the super resolution approach consists of decomposing it into $\mathbf{T}_d(\mathbf{a}) = \mathbf{T}_d(\mathbf{y}^+) - \mathbf{T}_d(\mathbf{y}^-)$, optimizing over $\mathbf{y}^+$ and $\mathbf{y}^-$, and then applying the "Toeplitz Prony" method to $\mathbf{T}_d(\mathbf{y}_*^+)$ and $\mathbf{T}_d(\mathbf{y}_*^-)$ at an optimal solution $(\mathbf{y}_*^+, \mathbf{y}_*^-)$ of (6.16).

Theorem 6.5. *For each $d \geq \eth$ the (complex) semidefinite program $(\boldsymbol{P}_{d,\eth})$ in (6.16) has an optimal solution $(\mathbf{y}^+, \mathbf{y}^-)$. In addition, if*

$$\mathrm{rank}(\mathbf{T}_d(\mathbf{y}^+)) = \mathrm{rank}(\mathbf{T}_{d-2}(\mathbf{y}^+)) \qquad (6.17)$$

$$\mathrm{rank}(\mathbf{T}_d(\mathbf{y}^-)) = \mathrm{rank}(\mathbf{T}_{d-2}(\mathbf{y}^-)) \qquad (6.18)$$

then there exist two Borel atomic measures ϕ^+ and ϕ^- on $\mathbb{T}^n$ such that:

$$y_\alpha^+ = \int_{\mathbb{T}^n} \mathbf{z}^\alpha \, d\phi^+(\mathbf{z}) \quad and \quad y_\alpha^- = \int_{\mathbb{T}^n} \mathbf{z}^\alpha \, d\phi^-(\mathbf{z}), \quad \forall \alpha \in \mathscr{A}_d'. \qquad (6.19)$$

The support of ϕ^+ (resp. ϕ^-) consists of $\mathrm{rank}(\mathbf{T}_d(\mathbf{y}^+))$ (resp. $\mathrm{rank}(\mathbf{T}_d(\mathbf{y}^-)))$ atoms on $\mathbb{T}^n$ which can be extracted by a numerical algebra routine.

In addition, if (6.17)-(6.18) hold for an optimal solution of $(\boldsymbol{P}_{d,\tilde{\eth}})$ with $\tilde{\eth}$ as in (6.15), then under the separation conditions of Theorem 6.4, the Borel measure $\phi^* := \phi^+ - \phi^-$ is the unique optimal solution of (6.6).

Proof. Consider a minimizing sequence $(\mathbf{y}^{+,\ell}, \mathbf{y}^{-,\ell})_{\ell \in \mathbb{N}}$ of (6.16). Since one minimizes $y_0^+ + y_0^-$ one has $y_0^{+,\ell} + y_0^{-,\ell} \leqslant y_0^{+,1} + y_0^{-,1} =: \rho$, for $\ell \geq 1$. The Toeplitz-like structure of $\mathbf{T}_d(\mathbf{y}^{+,\ell})$ and the psd constraint $\mathbf{T}_d(\mathbf{y}^{+,\ell}) \succeq 0$ imply $|y_\alpha^{+,\ell}| \leqslant \rho$ for all $\alpha \in \mathscr{A}_d'$; and similarly $|y_\alpha^{-,\ell}| \leqslant \rho$ for all $\alpha \in \mathscr{A}_d'$. Hence there is a subsequence (ℓ_k) and two vectors $\mathbf{y}^+ = (y_\alpha^+)_{\alpha \in \mathscr{A}_d'}$ and $\mathbf{y}^- = (y_\alpha^-)_{\alpha \in \mathscr{A}_d'}$, such that

$$\lim_{k \to \infty} \mathbf{y}^{+,\ell_k} = \mathbf{y}^+ \quad and \quad \lim_{k \to \infty} \mathbf{y}^{-,\ell_k} = \mathbf{y}^-.$$

In addition, from the above convergence it also follows that $(\mathbf{y}^+, \mathbf{y}^-)$ is a feasible solution of (6.16), hence an optimal solution of (6.16).

Next, in the univariate case, a Borel measure ϕ^+ and ϕ^- on $\mathbb{T}$ can always be extracted from the semidefinite positive Toeplitz matrices $\mathbf{T}_d(\mathbf{y}^+)$ and $\mathbf{T}_d(\mathbf{y}^-)$ respectively. This is true regardless of the rank conditions (6.17)-(6.18) and was proved in Iohvidov (1982)[p. 211]. In the multivariate case, $\mathbf{T}_d(\mathbf{y}^+)$ and $\mathbf{T}_d(\mathbf{y}^-)$ are Toeplitz-like matrices, and we may and will invoke a result from Josz and Molzahn (2018)[Theorem 5.2]. It implies that a Borel measure ϕ^+ (resp. ϕ^-) on $\mathbb{T}^n$ can be extracted from a multivariate semidefinite positive Toeplitz-like matrix $\mathbf{T}_d(\mathbf{y}^+)$ (resp. $\mathbf{T}_d(\mathbf{y}^-)$) if the rank condition (6.17) (resp. (6.18)) holds. Hence we have proved (6.19).

Finally recall that Theorem 6.4 is still valid with $\mathscr{A}'_0$ in lieu of $\mathscr{A}^1_0$ and therefore the last statement follows from Theorem 6.4 and the fact that (6.6) and (6.8) have same optimal value and an optimal solution (ϕ^+, ϕ^-) of (6.8) provides and optimal solution $\phi^* = \phi^+ - \phi^-$ of (6.6). □

6.3.1 *A rigorous compressed sensing LP approach*

In this section we take advantage of an important consequence of viewing sparse interpolation as a Super Resolution problem. Indeed when $\mathbf{z}_0$ is chosen as in (6.1) we *know* that the (unique) optimal solution ϕ^* of (6.6) is supported on the *a priori* fixed grid $(\exp(2i\pi k_1/N), \ldots, \exp(2i\pi k_n/N))$, where $0 \leq k_i \leq N$, $i = 1, \ldots, n$. That is, (6.6) is a *discrete* Super Resolution problem as described in Candès and Fernandez-Granda (2014). Therefore solving (6.6) is also equivalent to solving the LP:

$$\min_{\mathbf{x}} \{ \|\mathbf{x}\|_1 : \mathbf{A}\mathbf{x} = \mathbf{b} \}$$

where $\mathbf{x} \in \mathbb{R}^{[0,1,\ldots,N]^n}$. The matrix $\mathbf{A}$ has its columns indexed by $\beta \in [0, 1, \ldots, N]^n$ and its rows indexed by $\alpha \in \mathscr{A}^1_0$ (or $\mathscr{A}'_0$), while $\mathbf{b} = (b_\alpha)_{\alpha \in \mathscr{A}^1_0}$ (or $\mathbf{b} = (b_\alpha)_{\alpha \in \mathscr{A}'_0}$) is the vector of black-box evaluations at the points $(\mathbf{z}^\alpha_0)$, $\alpha \in \mathscr{A}^1_0$ (or $\alpha \in \mathscr{A}'_0$). So

$$\mathbf{A}(\alpha, \beta) = (\mathbf{z}^\beta_0)^\alpha = z^{\beta_1 \alpha_1}_{01} \cdots z^{\beta_n \alpha_n}_{0n}; \quad b_\alpha = g(\mathbf{z}^\alpha_0), \qquad (6.20)$$

for all $\alpha \in \mathscr{A}^1_0$ and $\beta \in [0, 1, \ldots, N]^n$.

Proposition 6.6. *Under the conditions of Theorem 6.4, the LP*

$$\rho = \min_{\mathbf{x}} \{ \|\mathbf{x}\|_1 : \mathbf{A}\mathbf{x} = \mathbf{b} \}$$

with $\mathbf{A}$ and $\mathbf{b}$ as in (6.20), has a unique optimal solution which is the vector of coefficients of the polynomial g^* of Theorem 6.4.

Proof. Let $\mathbf{x} \in \mathbb{R}^{[0,\ldots,N]^n}$ be an admissible solution of the LP with $\mathbf{A}$

and $\mathbf{b}$ as in (6.20). Let $g \in \mathbb{R}[\mathbf{x}]$, $\mathbf{z} \mapsto g(\mathbf{z}) := \sum_\beta g_\beta \, \mathbf{z}^\beta$, be the polynomial with vector of coefficients

$$g_\beta = \mathbf{x}_\beta \, (= x_{\beta_1}, \dots, x_{\beta_n}), \quad \forall \beta \in [0, \dots, N]^n.$$

By construction one has $g(\mathbf{z}_0^\alpha) = a_\alpha = g^*(\mathbf{z}_0^\alpha)$ for all $\alpha \in \mathscr{A}_1^1$, where g^* is as in Theorem 6.4. The Borel measures ν^+ and ν^- on $\mathbb{T}^n$ defined by

$$\nu^+ := \sum_{\beta \in [0, \dots, N]^n} \max[0, x_\beta] \, \delta_{\mathbf{z}_0^\beta}; \quad \nu^- := \sum_{\beta \in [0, \dots, N]^n} -\min[0, x_\beta] \, \delta_{\mathbf{z}_0^\beta},$$

are a feasible solution of (6.8) and the Borel signed measure $\nu := \nu^+ - \nu^-$ satisfies $\|\nu\|_{TV} = \|\nu^+\| + \|\nu^-\|$. Hence $\|\nu\|_{TV} \geq \|\phi^*\|_{TV}$ where ϕ^* is the optimal solution of (6.6). So the optimal value ρ of the LP satisfies $\rho \geq \|\phi^*\|_{TV}$. On the other hand with g^* as in Theorem 6.4, let

$$\mathbf{x}_\beta^* \, (= x_{\beta_1}^*, \dots, x_{\beta_n}^*) := g_\beta^*, \quad \forall \beta \in [0, \dots, N]^n.$$

Then $\|\mathbf{x}^*\|_1 = \|\phi^*\|_{TV} \leq \rho$ and so $\|\mathbf{x}^*\|_1 = \rho$, which proves that $\mathbf{x}^*$ is an optimal solution of the LP. Uniqueness follows from the uniqueness of solution to (6.6).   □

6.4   Numerical experiments

In polynomial interpolation, the number of evaluations $\eth$ is not given. Rather, we seek to recover a black-box polynomial using the *least* number of evaluations. Thus, one could set $\eth = 1$, then solve the hierarchy of SDPs (6.16) of order $d = 1, 2, \dots$. Next, set $\eth = 2$, and repeat. This leads to a hierarchy of hierarchies, which is costly from a computational perspective. Thus, we propose a *single* hierarchy where we choose to make all possible evaluations at each relaxation order d. Therefore we fix $d = \eth$ in (6.16) and let $\eth$ increase[2] to see if and when we recover the desired optimal measure (polynomial g^*) of Theorem 6.4.

---

[2] In the univariate case, the optimal value of $(\mathbf{P}_{d,\eth})$ in (6.16) does not increase with d when $d > \eth$. Indeed, for any optimal solution of $(\mathbf{P}_{\eth,\eth})$, there exists a representing signed measure $\phi = \phi^+ - \phi^-$ on the torus. However, one may not be able to extract this measure.

Separation of the support

Initially, super resolution was concerned with signal processing where the measurements are given and fixed and we have no influence on them. In contrast, in the super resolution formulation of polynomial interpolation problem, we can *choose* where we make the measurements, that is the points where we evaluate the black-box polynomial. This can have a strong influence on the separation condition which guarantees exact recovery of the signal (our black-box polynomial). We illustrate this phenomenon on the following example. Suppose that we are looking for the black-box polynomial

$$f(x) = 3x^{20} + x^{75} - 6x^{80} \qquad (6.21)$$

whose degree we assume to be less than or equal to 100. We consider such a high degree in order to well illustrate the notion of separation of the support. We first investigate two different ways of making evaluations and their impact on separation of the support, crucial for Super Resolution. Then we consider more realistic polynomials, limited to degree 10. So first let us evaluate the black-box polynomial at the points

$$(e^{\frac{2\pi i}{101}})^0, (e^{\frac{2\pi i}{101}})^1, (e^{\frac{2\pi i}{101}})^2, \ldots, (e^{\frac{2\pi i}{101}})^d \subset \mathbb{T} \qquad (6.22)$$

at step d of the SDP hierarchy (i.e. $(\mathbf{P}_{d,d})$ in (6.16)). The proximity of points on the torus is thus directly related to the proximity of the exponents of the polynomial. It can be seen in the left part of Figure 6.2 that some of the points on the torus are very close to each other.

Let us secondly evaluate in the black-box polynomial at the points

$$(e^i)^0, (e^i)^1, (e^i)^2, \ldots, (e^i)^d \subset \mathbb{T} \qquad (6.23)$$

at step d of the SDP hierarchy. The proximity of points on the torus is thus no longer related to the proximity of the exponents of the polynomial. It can be seen in the left part of Figure 6.3 that the points on the torus are nicely spread out. This is not guaranteed to be the case, but is expected to be true generically. In order to recover the black-box polynomial once a candidate atomic measure is computed, we form a table of the integers

Table 6.1   Evaluation at roots of unity $e^{\frac{2ki\pi}{101}}$

$d = \eth$	$\|\mu\|_{TV}$	#supp(μ)
0	2.0000	1
1	7.6618	2
2	8.1253	3
3	8.3655	5
4	8.7240	7
5	8.9882	9
6	9.3433	11
7	9.5837	13
8	9.7993	17
9	9.9436	19
10	9.9978	20
11	10.0000	3

Table 6.2   Evaluation at e^{ki}

$d = \eth$	$\|\mu\|_{TV}$	#supp(μ)
0	2.0000	1
1	8.7759	2
2	9.2803	3
3	10.0000	3

$k = 1, \ldots, d$ modulo 2π. For each atom, we consider its argument and find the closest value in the table, yielding an integer k, i.e. the power of the monomial associated to the atom. The coefficient of the monomial is given by the weight of the atom. In Table 6.1 and Table 6.2 are displayed the optimal value and the number of atoms of the optimal measure μ at each order d. Graphical illustrations of the solutions appear in Figure 6.2 and Figure 6.3. The dual polynomials in the right hand of the figures illustrate why a higher degree is needed when the points are closer.

Remark 6.7. Naive LP with evaluations at random points on the real line requires about 50 evaluations on this example, compared with the 4 evaluations with Super Resolution using multiple loops and in fact, the rigorous LP on the torus also requires 4 evaluations.

Additional numerical experiments

(1) **Generation of the examples:** We consider a random set of ten sparse polynomials with up to ten variables and up to degree ten (first

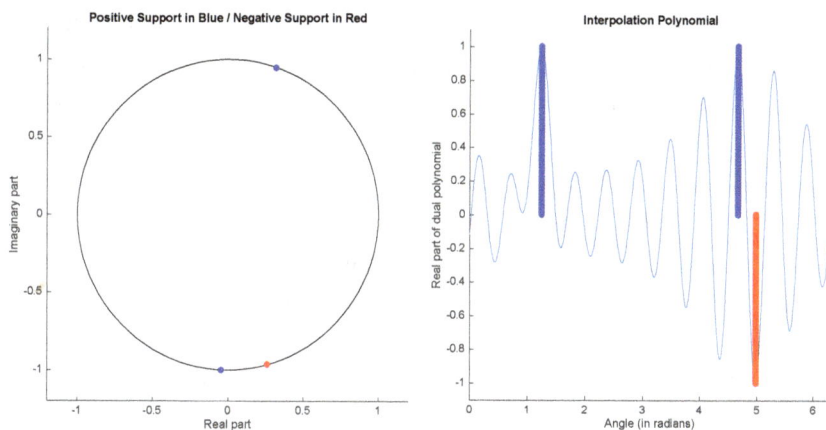

Fig. 6.2 Primal-dual solution of Super Resolution at order 11 (using single loop)

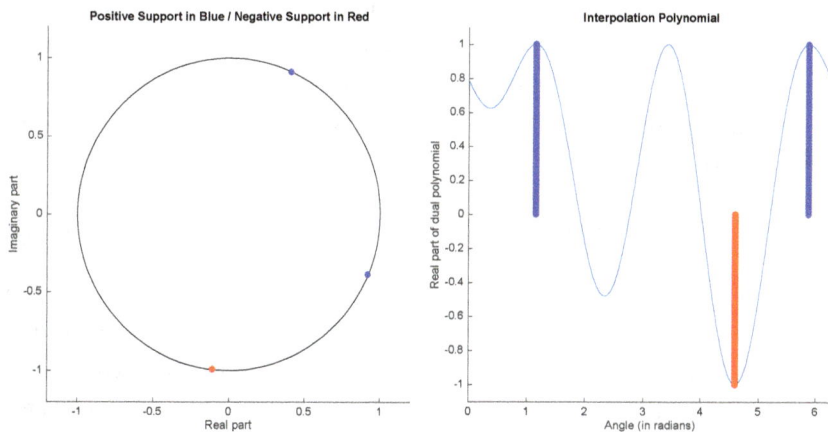

Fig. 6.3 Primal-dual solution of Super Resolution at order 3 (using multiple loops)

column of Table 6.3). For example, for a polynomial of n variables and k atoms, we generate the exponents β of the k monomials $\mathbf{x}^\beta$ randomly from $\mathbb{N}_d^n := \{\beta \in \mathbb{N}^n \mid \sum_{i=1}^n \beta_i \leqslant d\}$ and the associated non-zero coefficients g_β are drawn from a uniform distribution in the interval $[-10, 10]$.

(2) **Results in the noiseless case:** We detect the minimum number of

evaluations for each approach to recover the black-box polynomial in the noiseless case and report the results in Table 6.3. We use evaluations at the points $e^{i\,\alpha} = (e^{i\,\alpha_1}, \ldots, e^{i\,\alpha_n})$ with $\alpha = (\alpha_1, \ldots, \alpha_n) \in \mathbb{Z}^n$ up to a certain degree $\sum_{k=1}^{n} |\alpha_k| \leqslant d$. The corresponding number of evaluations and degree (d) are reported in the columns *Rigorous LP*, *Super Resolution*, and *Toeplitz Prony* of Table 6.3. In *Advanced T. Prony*, evaluations are made at different points. Thus, only the first three columns of Table 6.3 can be compared in presence of noise.

(3) **Results in the presence of noise:** For each of the ten polynomials we determine the maximum degree $d^{\max}$ for the evaluations $g(e^{i\alpha_1}, \ldots, e^{i\alpha_n})$ with $\alpha_1, \ldots, \alpha_n \in \mathbb{Z}$ and $\sum_{i=1}^{n} |\alpha_k| \leq d^{\max}$, in the *Rigorous LP* and *Super Resolution* approaches; results are displayed in Table 6.3. For example, for the first line of Table 6.3, that number is $d^{\max} := 2$ which corresponds to 3 evaluations in this univariate problem. As a result, we know that for these evaluations both approaches return the correct sparse polynomial. We then add uniform noise to those evaluations, i.e.

$$g(e^{i\,\alpha_1}, \ldots, e^{i\,\alpha_n}) + \epsilon_\alpha \, , \quad \epsilon_\alpha \in \mathbb{C}, \quad \mathfrak{Re}_\alpha, \mathfrak{Ie}_\alpha \in [-0.1, +0.1] \quad (6.24)$$

for all $|\alpha_1| + \cdots + |\alpha_n| \leqslant d^{\max}$ and $\alpha_1, \ldots, \alpha_n \in \mathbb{Z}$. Next, we run each approach ten times (with new noise every time) and report the average error in Table 6.4. The error is defined as the relative error in percentage of the output polynomial $\hat{g}(\mathbf{x}) = \sum_\alpha \hat{g}_\alpha \mathbf{x}^\alpha$ compared with the black-box polynomial $g(\mathbf{x}) = \sum_\alpha g_\alpha \mathbf{x}^\alpha$ using the 2 norm of the coefficients, i.e.

$$100 \times \frac{\sqrt{\sum_\alpha (\hat{g}_\alpha - g_\alpha)^2}}{\sqrt{\sum_\alpha g_\alpha^2}}. \quad (6.25)$$

Note that in *Rigorous LP* and *Super Resolution* the equality constraints associated with the evaluations are relaxed to inequalities, a functionality which is not possible in Prony method. This allows for more robustness. Precisely, in *Rigorous LP*, we replace $\mathbf{Ax} - \mathbf{b} = 0$ by $-0.1 \leqslant \mathfrak{R}(\mathbf{Ax} - \mathbf{b}) \leqslant 0.1$ and $-0.1 \leqslant \mathfrak{I}(\mathbf{Ax} - \mathbf{b}) \leqslant 0.1$, while in

Table 6.3   Minimum number of evaluations and degrees without noise; evaluations at the points $(e^{i\,\alpha_1}, \ldots, e^{i\,\alpha_n})$ for $|\alpha_1| + \cdots + |\alpha_n| \leqslant d$

Black-box Polynomial	Rigorous LP	Super-Resolution
$-1.2x^4 + 6.7x^7$	2 (1)	3 (2)
$2.3x^6 + 5.6x^3 - 1.5x^2$	4 (3)	5 (4)
$-2.1x^3 + 5.4x^2 - 2.0x + 6.2x^5 - 5.2$	5 (4)	6 (5)
$0.8x_1x_2 - x_1x_2^2$	19 (3)	31 (4)
$-5.8x_1^2x_2^2 - 8.2x_1^2x_2^3 + 5.5x_1^3x_2 + 1.1$	10 (2)	19 (3)
$-7.2x_1x_2^2 + 1.8x_1^3x_2^2 + 2.6x_1^4x_2^5 + 6.2x_1x_2^5 + 2.5x_1$	10 (2)	19 (3)
$-3.5 + 8.1x_1^3x_2x_3$	7 (1)	28 (2)
$-1.2x_1^2x_2^2x_3^3 + 7.3x_1^2x_2 - 2.4x_2$	28 (2)	28 (2)
$-6.1x_1^2x_5 + 2.5x_2x_4 + 4.8x_3$	136 (2)	136 (2)
$2.9x_2x_3x_9^4x_{10} - 5.6x_1x_4^2x_7 - 4.1x_3x_5x_6^3x_8$	N. A.	1595 (2)

Table 6.4   Relative error with uniform noise for the real and imaginary parts on the measurements; evaluations at the points $(e^{i\alpha_1}, \ldots, e^{i\alpha_n})$ for $|\alpha_1| + \cdots + |\alpha_n| \leqslant d$

Black-box Polynomial	Rigorous LP	Super-Resolution
$-1.2x^4 + 6.7x^7$	4.18%	1.58%
$2.3x^6 + 5.6x^3 - 1.5x^2$	1.94%	1.81%
$-2.1x^3 + 5.4x^2 - 2.0x + 6.2x^5 - 5.2$	1.47%	1.40%
$0.8x_1x_2 - x_1x_2^2$	3.23%	4.84%
$-5.8x_1^2x_2^2 - 8.2x_1^2x_2^3 + 5.5x_1^3x_2 + 1.1$	1.13%	0.87%
$-7.2x_1x_2^2 + 1.8x_1^3x_2^2 + 2.6x_1^4x_2^5 + 6.2x_1x_2^5 + 2.5x_1$	1.23%	1.08%
$-3.5 + 8.1x_1^3x_2x_3$	0.79%	0.70%
$-1.2x_1^2x_2^2x_3^3 + 7.3x_1^2x_2 - 2.4x_2$	2.19%	1.03%
$-6.1x_1^2x_5 + 2.5x_2x_4 + 4.8x_3$	0.94%	1.15%
$2.9x_2x_3x_9^4x_{10} - 5.6x_1x_4^2x_7 - 4.1x_3x_5x_6^3x_8$	N. A.	0.47%

Super Resolution we use a 2-norm ball of radius $0.1 \times \sqrt{2}$ (similar to the technique employed in Candès and Fernandez-Granda (2014)).

6.5   Notes and sources

This chapter is essentially taken from Josz et al. (2019) in which the rigorous LP and Super Resolution approaches are also compared with two variants of Prony's method.

Bibliography

Emmanuel J. Candès, E. J. (2008). *The restricted isometry property and its implications for compressed sensing*, C.R. Acad. Sci. Paris Ser. I **346**, pp. 589–592.

Candès, E. and Fernandez-Granda, C. (2013). *Super-Resolution from Noisy Data*, J. Fourier Anal. Appl. **19**, pp. 1229–1254.

Candès, E. and Fernandez-Granda, C. (2014). *Towards a Mathematical Theory of Super-resolution*, Com. Pure Appl. Math. **67**, pp. 906–956.

I. S. Iohvidov, I. S. (1985). *Hankel and Toeplitz Matrices and Forms: Algebraic Theory*, (Birkhäuser Verlag, Boston).

Josz, C., and Molzahn, D. (2018). *Lasserre Hierarchy for Large Scale Polynomial Optimization in Real and Complex Variables*, SIAM J. Optim. **28**, pp. 1017–1048.

Josz, C., Lasserre, J. B., Mourrain, B. (2019). *Sparse polynomial interpolation: Compressed sensing, super resolution, or Prony?*, Adv. Math. Computation **45**, pp. 1401–1437.

Kunis, S., Thomas, P., Römer, T., and von der Ohe, U. (2016). *A multivariate generalization of Prony's method*, Lin. Alg. Appl. **490**, pp. 31–47.

de Baron de Prony, G. R. (1795). *Essai experimental et analytique: Sur les lois de la dilatabilité de fluides élastique et sur celles de la force expansive de la vapeur de l'alcool, à différentes temperatures*, J. Ecole Polytechnique **1**, pp. 24–76.

Chapter 7

Representation of (Probabilistic) Chance-Constraints

We describe how to apply the Moment-SOS hierarchy to obtain a representation of chance-constraints with strong asymptotic properties.

7.1 Introduction

Consider the following general framework for decision under uncertainty: Let $\mathbf{x} \in \mathbf{X} \subset \mathbb{R}^n$ be a decision variable while $\boldsymbol{\omega} \in \mathbb{R}^p$ is a *disturbance* (or *noise*) parameter whose distribution μ (with support $\boldsymbol{\Omega} \subset \mathbb{R}^p$) is known, i.e., its list of moments $\mu_\beta := \int_{\boldsymbol{\Omega}} \boldsymbol{\omega}^\beta \, d\mu(\boldsymbol{\omega})$, $\beta \in \mathbb{N}^p$, is available in closed form or numerically. Both $\mathbf{x}$ and $\boldsymbol{\omega}$ are linked by constraints of the form $(\mathbf{x}, \boldsymbol{\omega}) \in \mathbf{K} \subset \mathbf{X} \times \boldsymbol{\Omega}$, where

$$\mathbf{K} = \{ (\mathbf{x}, \boldsymbol{\omega}) : g_j(\mathbf{x}, \boldsymbol{\omega}) \geq 0, \quad j = 1, \ldots, m \}, \tag{7.1}$$

for some polynomials $(g_j) \subset \mathbb{R}[\mathbf{x}, \boldsymbol{\omega}]$. Let $\varepsilon \in (0,1)$ be fixed. The goal is to approximate as closely as possible, the set:

$$\mathbf{X}_\varepsilon^* = \{ \mathbf{x} \in \mathbf{X} : \mathrm{Prob}((\mathbf{x}, \boldsymbol{\omega}) \in \mathbf{K}) \geq 1 - \varepsilon \}. \tag{7.2}$$

One provides a hierarchy of outer approximations of $\mathbf{X}_\varepsilon^*$ in the form $\mathbf{X}_\varepsilon^d := \{ \mathbf{x} \in \mathbf{X} : h_d(\mathbf{x}) \geq 0 \}$, $d \in \mathbb{N}$, (with h_d a polynomial of degree at most $2d$) and with the strong asymptotic property:

$$\lim_{d \to \infty} \lambda(\mathbf{X}_\varepsilon^d \setminus \mathbf{X}_\varepsilon^*) \to 0, \tag{7.3}$$

where λ is the Lebesgue measure on $\mathbf{X}$.

The same methodology applied to the set $\{\mathbf{x} \in \mathbf{X} : \mathrm{Prob}((\mathbf{x}, \boldsymbol{\omega}) \in \mathbf{K}) \leq \varepsilon\}$ yields a hierarchy of *inner* approximations to $\mathbf{X}_\varepsilon^*$ in (7.2), in the form $\hat{\mathbf{X}}_\varepsilon^d := \{\mathbf{x} \in \mathbf{X} : q_d(\mathbf{x}) \geq 0\}$, $d \in \mathbb{N}$, for some polynomials $(q_d) \subset \mathbb{R}[\mathbf{x}]$ with the strong asymptotic property $\lim_{d \to \infty} \lambda(\mathbf{X}_\varepsilon^* \setminus \hat{\mathbf{X}}_\varepsilon^d) \to 0$.

Methodology. To obtain the polynomials $(h_d) \subset \mathbb{R}[\mathbf{x}]$:

- We first introduce an infinite-dimensional linear optimization problem (LP) on an appropriate space of Borel measures, whose optimal value ρ satisfies:

$$\hat{\rho} = \int_{\mathbf{X}} \rho(\mathbf{x}) \, d\lambda(\mathbf{x}) \quad \text{with} \quad \rho(\mathbf{x}) := \mathrm{Prob}((\mathbf{x}, \boldsymbol{\omega}) \in \mathbf{K}), \quad \mathbf{x} \in \mathbf{X},$$

- We then apply the Moment-SOS hierarchy for solving this LP. At step d of this hierarchy one solves a semidefinite relaxation of this LP whose dual provides a polynomial h_d of degree at most d with the property (i) $h_d(\mathbf{x}) \geq \rho(\mathbf{x})$ for all $\mathbf{x} \in \mathbf{X}$ and (ii), $\|h_d - \rho\|_{L_1(\mathbf{X}, \lambda)} \to 0$ as d increases. The desired result (7.3) follows from this convergence.

7.2 An infinite-dimensional LP

Let $\mathbf{X} \subset \mathbb{R}^n$ be a compact basic semi-algebraic set and let λ be the Lebesgue measure on $\mathbf{X}$. With $\mathbf{K} \subset \mathbf{X} \times \boldsymbol{\Omega}$ as in (7.1), and for every $\mathbf{x} \in \mathbf{X}$, let $\mathbf{K_x} \subset \boldsymbol{\Omega}$ be the (possibly empty) set:

$$\mathbf{K_x} := \{\boldsymbol{\omega} \in \boldsymbol{\Omega} : (\mathbf{x}, \boldsymbol{\omega}) \in \mathbf{K}\}, \quad \mathbf{x} \in \mathbf{X},$$

and let $\lambda \otimes \mu$ be the product measure on $\mathbf{X} \times \boldsymbol{\Omega}$, i.e.,

$$\lambda \otimes \mu(A \times B) = \lambda(A) \cdot \mu(B), \quad \forall A \in \mathcal{B}(\mathbf{X}), \, B \in \mathcal{B}(\boldsymbol{\Omega}).$$

Consider the infinite-dimensional linear program (LP):

$$\rho = \sup_{\phi \in \mathcal{M}(\mathbf{K})_+} \{\phi(\mathbf{K}) : \phi \leq \lambda \otimes \mu\}. \qquad (7.4)$$

Theorem 7.1. *The unique optimal solution ϕ^* of (7.4) is the restriction of $\lambda \otimes \mu$ to $\mathbf{K}$ and the optimal value ρ of (7.4) satisfies*

$$\hat{\rho} = \phi^*(\mathbf{K}) = \int_{\mathbf{K}} d(\lambda \otimes \mu) = \int_{\mathbf{X}} \mu(\mathbf{K_x}) \lambda(dx) = \int_{\mathbf{X}} \rho(\mathbf{x}) \lambda(dx). \quad (7.5)$$

Proof. By Theorem 3.4 (with $\lambda \otimes \mu$ instead of μ) $d\phi^* = 1_{\mathbf{K}}((\mathbf{x}, \boldsymbol{\omega})) \lambda \otimes \mu(d(\mathbf{x}, \boldsymbol{\omega}))$. Next, by Fubini-Tonelli's Theorem

$$\rho = \int_{\mathbf{X} \times \Omega} 1_{\mathbf{K}}((\mathbf{x}, \boldsymbol{\omega})) \lambda \otimes \mu(d(\mathbf{x}, \boldsymbol{\omega})) = \int_{\mathbf{X}} \underbrace{\int_{\Omega} 1_{\mathbf{K}}((\mathbf{x}, \boldsymbol{\omega})) \mu(d\boldsymbol{\omega})}_{\mu(\mathbf{K_x})} \lambda(dx)$$

$$= \int_{\mathbf{X}} \mu(\mathbf{K_x}) \lambda(dx),$$

and the function $\rho : \mathbf{X} \mapsto \mathbb{R}_+$, $\mathbf{x} \mapsto \rho(\mathbf{x}) := \mu(\mathbf{K_x})$ is measurable. □

7.3 The Moment-SOS hierarchy

7.3.1 *Outer approximations*

With $\mathbf{K}$ as in (7.1), let $d_j = \lceil \deg(g_j)/2 \rceil$ for all j, and consider the following hierarchy of semidefinite programs, indexed by $d \in \mathbb{N}$:

$$\rho_d = \sup_{\mathbf{y}, \mathbf{z}} \{ y_0 : y_{\alpha, \beta} + z_{\alpha, \beta} = \lambda_\alpha \cdot \mu_\beta, \quad \forall (\alpha, \beta) \in \mathbb{N}_{2d}^{n+p}$$
$$\mathbf{M}_d(\mathbf{y}), \mathbf{M}_d(\mathbf{z}) \succeq 0 \quad (7.6)$$
$$\mathbf{M}_{d-d_j}(g_j \, \mathbf{y}) \succeq 0, \quad j = 1, \ldots, m \},$$

and of course $\rho_d \geq \rho_{d+1} \geq \rho$ for all d. The dual of (7.6) is the semidefinite program:

$$\rho_d^* = \inf_{p \in \mathbb{R}[\mathbf{x}, \boldsymbol{\omega}]_{2d}} \{ \int_{\mathbf{X} \times \Omega} p(\mathbf{x}, \boldsymbol{\omega}) \lambda \otimes \mu(d(\mathbf{x}, \boldsymbol{\omega}))$$
$$\text{s.t. } p(\mathbf{x}, \boldsymbol{\omega}) \geq 1, \quad \forall (\mathbf{x}, \boldsymbol{\omega}) \in \mathbf{K} \quad (7.7)$$
$$p \text{ is SOS} \}.$$

Again as $\mathbf{K}$ is compact, for technical reasons (but with no loss of generality) we may and will assume that in the definition (7.1) of $\mathbf{K}$, $g_1(\mathbf{x}) = M - \|\mathbf{x}\|^2$ for some $M > 0$.

Theorem 7.2. *Let* $\mathbf{K}$ *and* $(\mathbf{X} \times \mathbf{\Omega}) \setminus \mathbf{K}$ *be with nonempty interior. There is no duality gap between (7.6) and its dual (7.7), i.e.,* $\rho_d = \rho_d^*$ *for all d. In addition (7.7) has an optimal solution* $p_d^* \in \mathbb{R}[\mathbf{x}, \boldsymbol{\omega}]_{2d}$ *such that*

$$\rho_d = \rho_d^* = \int_{\mathbf{X} \times \mathbf{\Omega}} p_d^*(\mathbf{x}, \boldsymbol{\omega}) \, \lambda \otimes \mu(d(\mathbf{x}, \boldsymbol{\omega})).$$

Define $h_d^* \in \mathbb{R}[\mathbf{x}]_{2d}$ *to be:*

$$\mathbf{x} \mapsto h_d^*(\mathbf{x}) := \int_{\mathbf{\Omega}} p_d^*(\mathbf{x}, \boldsymbol{\omega}) \mu(d\boldsymbol{\omega}), \quad \mathbf{x} \in \mathbb{R}^n.$$

Then $h_d^*(\mathbf{x}) \geq \rho(\mathbf{x})$ *for all* $\mathbf{x} \in \mathbf{X}$ *and*

$$\rho_d = \int_{\mathbf{X}} h_d^*(\mathbf{x}) \, \lambda(d\mathbf{x}) \to \hat{\rho} = \int_{\mathbf{X}} \rho(\mathbf{x}) \, \lambda(d\mathbf{x})$$

as $d \to \infty$. *Moreover*

$$\lim_{d \to \infty} \|h_d^* - \rho\|_{L_1(\mathbf{X}, \lambda)} = 0. \tag{7.8}$$

Proof. That $\rho_d = \rho_d^*$ is because Slater's condition holds for (7.6). Indeed let $\mathbf{y}^*$ be the moments of ϕ^* in Theorem 7.1 and $\mathbf{z}^*$ be the moments of $\lambda \otimes \mu - \phi^*$ (on $(\mathbf{X} \times \mathbf{\Omega}) \setminus \mathbf{K}$). Then as $\mathbf{K}$ has nonempty interior, $\mathbf{M}_d(\mathbf{y}^*) \succ 0$ and $\mathbf{M}_d(g_j \, \mathbf{y}^*) \succ 0$ for all d. Similarly as $(\mathbf{X} \times \mathbf{\Omega}) \setminus \mathbf{K}$ has nonempty interior, $\mathbf{M}_d(\mathbf{z}^*) \succ 0$. Moreover since the optimal value ρ_d is finite for all d this implies that (7.7) has an optimal solution $p_d^* \in \mathbb{R}[\mathbf{x}, \boldsymbol{\omega}]_{2d}$. Therefore:

$$\rho_d = \int_{\mathbf{X} \times \mathbf{\Omega}} p_d^*(\mathbf{x}, \boldsymbol{\omega}) \, \lambda \otimes \mu(d(\mathbf{x}, \boldsymbol{\omega}))$$

$$= \int_{\mathbf{X}} \underbrace{\left(\int_{\mathbf{\Omega}} p_d^*(\mathbf{x}, \boldsymbol{\omega}) \, \mu(d\boldsymbol{\omega}) \right)}_{h_d^*(\mathbf{x}) \geq \mu(\mathbf{K}_{\mathbf{x}})} \lambda(d\mathbf{x}) = \int_{\mathbf{X}} h_d^*(\mathbf{x}) \, \lambda(d\mathbf{x})$$

where $h_d^*(\mathbf{x}) \geq \mu(\mathbf{K}_{\mathbf{x}}) = \rho(\mathbf{x})$ follows from $p_d^* \geq 1$ on $\mathbf{K}$ and $\mu(\mathbf{K}_{\mathbf{x}}) = 0$ whenever $\mathbf{K}_{\mathbf{x}} = \emptyset$. Next, the convergence $\lim_{d \to \infty} \rho_d = \hat{\rho}$ follows from Theorem 3.2(c) (with μ replaced with $\lambda \otimes \mu$). Combining yields

$$0 = \lim_{d \to \infty} \rho_d - \hat{\rho} = \lim_{d \to \infty} \int_{\mathbf{X}} \underbrace{(h_d^*(\mathbf{x}) - \rho(\mathbf{x}))}_{\geq 0} \lambda(d\mathbf{x}) = \|h_d^* - \rho\|_{L_1(\mathbf{X}, \lambda)}.$$

$\square$

As a consequence:

Corollary 7.3. *With $\varepsilon \in (0,1)$ fixed, arbitrary, let $\mathbf{X}_\varepsilon^*$ be as in (7.2). Under the same hypotheses, let $h_d^* \in \mathbb{R}[\mathbf{x}]_{2d}$ be as in Theorem 7.2 and define $\mathbf{X}_\varepsilon^d := \{\mathbf{x} \in \mathbf{X} : h_d^*(\mathbf{x}) \geq 1 - \varepsilon\}$. Then:*

$$\mathbf{X}_\varepsilon^* \subset \mathbf{X}_\varepsilon^d, \quad \forall d \quad and \quad \lim_{d\to\infty} \lambda(\mathbf{X}_\varepsilon^d \setminus \mathbf{X}_\varepsilon^*) = 0. \tag{7.9}$$

Proof. From Theorem 7.2, the $L_1(\mathbf{X}, \lambda)$ convergence (7.8) implies convergence in λ-measure, that is, for every $\varepsilon > 0$, $\lambda(\{\mathbf{x} : |h_d(\mathbf{x}) - \rho(\mathbf{x})| > \varepsilon\}) \to 0$ as $d \to \infty$. Next, observe that $\mathbf{X} \setminus \mathbf{X}_\varepsilon^* = \bigcup_{\ell=1}^\infty \{\mathbf{x} \in \mathbf{X} : \mu(\mathbf{K}_\mathbf{x}) < 1 - \varepsilon - 1/\ell\}$ and therefore

$$\lambda(\mathbf{X} \setminus \mathbf{X}_\varepsilon^*) = \lim_{\ell\to\infty} \lambda(\underbrace{\{\mathbf{x} \in \mathbf{X} : \mu(\mathbf{K}_\mathbf{x}) < 1 - \varepsilon - 1/\ell\}}_{R_\ell}).$$

For each $\ell = 1, \ldots$, write

$$\lambda(R_\ell) = \lambda(R_\ell \cap \{\mathbf{x} \in \mathbf{X} : h_d^*(\mathbf{x}) < 1 - \varepsilon\}) + \lambda(R_\ell \cap \mathbf{X}_\varepsilon^d).$$

By the convergence in λ-measure, $\lim_{d\to\infty} \lambda(R_\ell \cap \mathbf{X}_\varepsilon^d) = 0$ as $d \to \infty$, and so

$$\begin{aligned}
\lambda(R_\ell) &= \lim_{d\to\infty} \lambda(R_\ell \cap \{\mathbf{x} \in \mathbf{X} : h_d^*(\mathbf{x}) < 1 - \varepsilon\}) \\
&\leq \lim_{d\to\infty} \lambda(\{\mathbf{x} \in \mathbf{X} : h_d^*(\mathbf{x}) < 1 - \varepsilon\}) \leq \lambda(\mathbf{X} \setminus \mathbf{X}_\varepsilon^*).
\end{aligned}$$

As this is true for every ℓ and $\lim_{\ell\to\infty} \lambda(R_\ell) = \lambda(\mathbf{X} \setminus \mathbf{X}_\varepsilon^*)$, we conclude that $\lim_{d\to\infty} \lambda(\{\mathbf{x} \in \mathbf{X} : h_d^*(\mathbf{x}) < 1 - \varepsilon\}) = \lambda(\mathbf{X} \setminus \mathbf{X}_\varepsilon^*)$, which in turn yields the desired result (7.9). □

7.3.2 Inner approximations

Assume that $\lambda \otimes \mu(\partial \mathbf{K}) = 0$ (which is the case if for instance μ has a density with respect to the Lebesgue measure, strictly positive on Ω). Then clearly, letting $\mathbf{K}^c := (\mathbf{X} \times \Omega) \setminus \mathbf{K}$, the same methodology now applied to a chance-constraint of the form

$$\text{Prob}((\mathbf{x}, \boldsymbol{\omega}) \in \mathbf{K}^c) < \varepsilon$$

would provide a converging sequence of *inner* approximations of the set $\mathbf{X}_\epsilon^* := \{\mathbf{x} \in \mathbf{X} : \mathrm{Prob}((\mathbf{x}, \boldsymbol{\omega}) \in \mathbf{K}) \geq 1 - \varepsilon\}$. To do so we write $\mathbf{K}^c$ as a finite union $\bigcup_i \mathbf{K}_i^c$ of basic semi-algebraic sets $\mathbf{K}_i^c$ (whose $\mu \otimes \lambda$ measure of their overlap is zero) and then apply the above methodology to each $\mathbf{K}_i^c$, and then sum-up over i.

7.3.3   *Accelerating convergence*

To compute outer approximations we need solve the semidefinite program (7.6) which (in a different context) is in the vein of the semidefinite program (3.6) for computing upper bounds on the volume of a semi-algebraic set. We have seen that the convergence the latter upper bounds as $d \to \infty$ is expected to be slow (because of a Gibbs phenomenon in solving the dual of (3.6)). However we have described a way to accelerate significantly this convergence. It consists in adding constraints in the relaxation (3.6) which come from properties of the optimal solution of (3.4), obtained from Stokes' Theorem.

A similar procedure can be applied in the present context of chance-constraints. Assume that μ is the Lebesgue measure on $\boldsymbol{\Omega}$ scaled to be a probability measure (but the same idea works if $d\mu = h(\boldsymbol{\omega})d\boldsymbol{\omega}$, or if $d\mu = v(\mathbf{x})\exp(h(\boldsymbol{\omega}))d\boldsymbol{\omega}$ for some polynomial h, v). Then again we include additional constraints on the moments $\mathbf{y}$ in (7.6) coming from additional properties of the optimal solution ϕ^* of (7.4). Again these additional properties are coming from Stokes' formula but now for integrals on $\mathbf{K_x}$, then integrated over $\mathbf{X}$.

Let $f \in \mathbb{R}[\mathbf{x}, \boldsymbol{\omega}]$ be the polynomial $(\mathbf{x}, \boldsymbol{\omega}) \mapsto f(\mathbf{x}, \boldsymbol{\omega}) = \prod_j g_j(\mathbf{x}, \boldsymbol{\omega})$. For each fixed $\mathbf{x} \in \mathbf{K_x}$, the polynomial $\boldsymbol{\omega} \mapsto f(\mathbf{x}, \boldsymbol{\omega})$ vanishes on the boundary $\partial \mathbf{K_x}$ of $\mathbf{K_x}$. Therefore for each $\beta \in \mathbb{N}^p$, Stokes' Theorem 1.10 states:

$$\int_{\mathbf{K_x}} \frac{\partial(\boldsymbol{\omega}^\beta f(\mathbf{x}, \boldsymbol{\omega}))}{\partial \omega_i} \, d\mu(\boldsymbol{\omega}) = 0, \quad \forall i = 1, \dots, n. \qquad (7.10)$$

So for each $(\alpha, \beta) \in \mathbb{N}^n \times \mathbb{N}^p$, define the polynomial $\theta_{i,\alpha,\beta} \in \mathbb{R}[\mathbf{x}, \boldsymbol{\omega}]$ of degree $d_{i,\alpha,\beta}$, by:

$$(\mathbf{x}, \boldsymbol{\omega}) \mapsto \theta_{i,\alpha,\beta}(\mathbf{x}, \boldsymbol{\omega}) := \mathbf{x}^\alpha \frac{\partial(\boldsymbol{\omega}^\beta f(\mathbf{x}, \boldsymbol{\omega}))}{\partial \omega_i}, \quad i = 1, \dots, n.$$

Then from (7.10), integrating with respect to λ yields:

$$\int_{\mathbf{X}} \int_{\mathbf{K_x}} \theta_{i,\alpha,\beta}(\mathbf{x},\boldsymbol{\omega})\,\mu(d\boldsymbol{\omega})\,\lambda(dx) = 0.$$

Equivalently, in view of what is ϕ^* in Theorem 7.1,

$$\int_{\mathbf{K}} \theta_{i,\alpha,\beta}(\mathbf{x},\boldsymbol{\omega})\,d\phi^*(\mathbf{x},\boldsymbol{\omega}) = 0.$$

Therefore in (7.6) we may include the additional moments constraints

$$L_{\mathbf{y}}(\theta_{i,\alpha,\beta}(\mathbf{x},\boldsymbol{\omega})) = 0, \quad \forall(\alpha,\beta) \in \mathbb{N}^{n+p};\; d_{i,\alpha,\beta} \leq 2d. \tag{7.11}$$

That is, we now consider the hierarchy of semidefinite programs indexed by $d \in \mathbb{N}$:

$$\begin{aligned}
\hat{\rho}_d = \sup_{\mathbf{y},\mathbf{z}} \{\, y_0 : y_{\alpha,\beta} + z_{\alpha,\beta} &= \lambda_\alpha \cdot \mu_\beta, \quad \forall\,(\alpha,\beta) \in \mathbb{N}^{n+p}_{2d} \\
\mathbf{M}_d(\mathbf{y}), \mathbf{M}_d(\mathbf{z}) &\succeq 0 \\
\mathbf{M}_{d-d_j}(g_j\,\mathbf{y}) &\succeq 0, \quad j = 1,\ldots,m \\
L_{\mathbf{y}}(\theta_{i,\alpha,\beta}(\mathbf{x},\boldsymbol{\omega})) &= 0, \quad d_{i,\alpha,\beta} \leq 2d\,\},
\end{aligned} \tag{7.12}$$

which satisfies $\rho \leq \hat{\rho}_d \leq \rho_d$ for all d.

7.3.4 *Numerical examples*

For illustration purposes we have considered simple small dimensional examples for which the function $\mathbf{x} \mapsto \rho(\mathbf{x}) := \mu(\mathbf{K_x})$ has a closed form expression, so that we can compare the set $\mathbf{X}^*_\varepsilon$ with its approximations $\mathbf{X}^d_\varepsilon$, $d \in \mathbb{N}$, obtained in Corollary 7.3 (with and without using Stokes constraints (7.11)).

Example 7.1. $\mathbf{X} = [-1,1]$, $\boldsymbol{\Omega} = [0,1]$ and $\mathbf{K} = \{(x,\omega) : 1 - x^2/0.81 - \omega^2/1.44 \geq 0\}$. λ and μ are the Lebesgue measure. In this case $\mathbf{X}^*_\varepsilon = [a_\varepsilon, b_\varepsilon]$, $\mathbf{X}^d_\varepsilon = [a^d_\varepsilon, b^d_\varepsilon] \subset [-1,1]$. In Table 7.1 we display the relative error $(b^d_\varepsilon - b_\varepsilon + a_\varepsilon - a^d_\varepsilon)/(b^d_\varepsilon - a^d_\varepsilon)$ for different values of ε and d, with and without Stokes constraints (7.11). The results indicate that adding Stokes constraints help a lot. With relatively few moments $2d \leq 16$ one obtains good approximations.

Table 7.1   Example 7.1; $n = 1$: with and without Stokes

	$d = 4$	$d = 4$ (Stokes)	$d = 8$	$d = 8$ (Stokes)
$\varepsilon = 0.75$	13%	3.8%	6.3%	0.7%
$\varepsilon = 0.50$	16.6%	2.1%	9.8%	0.2%
$\varepsilon = 0.25$	26.6%	6.5%	18.5%	1.0%
$\varepsilon = 0.00$	44.7%	22.7%	31.2%	8.2%

Example 7.2. $\mathbf{X} = [-1, 1]$, $\mathbf{\Omega} = [0, 1]$ and $\mathbf{K} = \{(x, \omega) : 1 - x^2 - \frac{\omega^2}{2} \geq 0; \frac{x^2}{2} + \omega^2 - \frac{1}{4} \geq 0\}$. λ and μ are the Lebesgue measure on $\mathbf{X}$ and $\mathbf{\Omega}$ respectively. In this case, when $\varepsilon < 0.4$, the set $\mathbf{X}_\varepsilon^*$ is the union of two disconnected intervals, hence more difficult to approximate. As in Example 7.1, in Table 7.2 we display the relative error for different values of ε and d, with and without Stokes constraints (7.11). Again, the results indicate that adding the Stokes constraints help a lot. With relatively few moments $2d \leq 20$ one obtains good approximations. For instance, $\mathbf{X}_{0.1}^* = [-0.7714, -0.3082] \cup [0.3082, 0.7714]$ while one obtains $\mathbf{X}_{0.1}^* \subset \mathbf{X}_{0.1}^{10} = [-0.7985, -0.26] \cup [0.26, 0.7985]$ with Stokes and the larger set $\mathbf{X}_{0.1}^{10} = [-0.8673, -0.1881] \cup [0.1881, 0.8673]$ without Stokes constraints.

Table 7.2   Example 7.2, $n = 1$; with and without Stokes

	$d = 7$	$d = 7$ (Stokes)	$d = 10$	$d = 10$ (Stokes)
$\varepsilon = 0.7$	2.3%	0.4%	2.3%	0.2%
$\varepsilon = 0.4$	30%	10%	26%	5.5%
$\varepsilon = 0.1$	52%	24%	46%	16%

Example 7.3. $\mathbf{X} = [-1, 1]^2$, $\mathbf{\Omega} = [0, 1]$ and $\mathbf{K} = \{(x, \omega) : 1 - x_1^2 - x_2^2 - \omega^2 \geq 0\}$. λ and μ are the Lebesgue measure on $\mathbf{X}$ and $\mathbf{\Omega}$ respectively. For this two-dimensional example (in $\mathbf{x}$) we have plotted the boundary of the set $\mathbf{X}_\varepsilon^*$ (inner approximate circle, solid line in black). The curve in the middle (red dashed line) (resp. outer, blue dashed line) is the boundary of the approximation $\mathbf{X}_\varepsilon^d$ computed with Stokes constraints (7.11) (resp. without Stokes constraints). For $\epsilon = 0.01$ and $d = 10$ the results are plotted in Figure 7.1(left), and for $\varepsilon = 0.05$ and $d = 10$ in Figure 7.1(right).

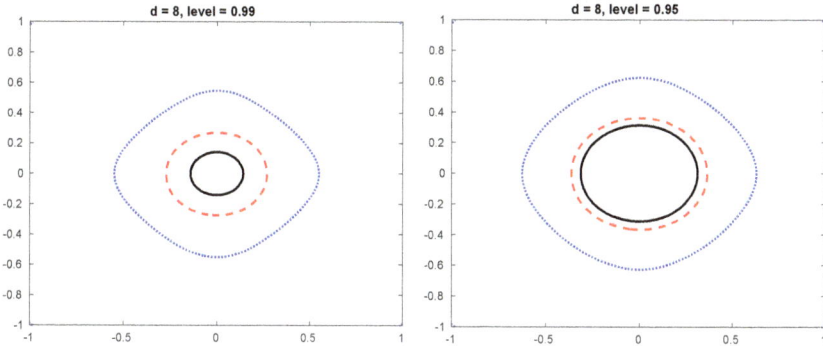

Fig. 7.1   Boundary of $\mathbf{X}^*_\epsilon$ and $\mathbf{X}^d_\epsilon$ with and without Stokes

7.4   Concluding remarks

We have shown that the Moment-SOS hierarchy can be used to provide outer and inner approximations of sets defined by chance-constraints of the form $\text{Prob}_\mu((\mathbf{x}, \boldsymbol{\omega}) \in \mathbf{K}) \geq 1 - \varepsilon$. We have assumed that the support $\boldsymbol{\Omega}$ of the distribution μ is compact but in fact one may also consider *any* probability measure μ, provided that (i) one knows all its moments, and (ii) if its support $\boldsymbol{\Omega}$ is not compact then μ should satisfy Carleman's condition (1.11). For instance μ can be the Gaussian measure on $\boldsymbol{\Omega} = \mathbb{R}^n$ or the exponential measure on $\boldsymbol{\Omega} = \mathbb{R}^n_+$ and the set $\mathbf{K_x}$, $\mathbf{x} \in \mathbf{X}$, can be non-compact. In such a case one may also accelerate the associated semidefinite relaxation (7.6) by including additional constraints derived from Stokes' Theorem applied to the optimal solution of (7.4). The latter Stokes' theorem is an extension of the standard one to handle such non-compact sets $\mathbf{K_x}$. For more details the interested reader is referred to Lasserre (2017a) and Lasserre (2017b).

7.5   Notes and sources

This chapter is essentially from Lasserre (2017a). Approximations of sets like $\mathbf{X}^*_\varepsilon$ in (7.2) are particularly useful for optimization and control problems with chance-constraints; for instance problems of the form:

$$\min \{ f(\mathbf{x}) : \mathbf{x} \in C; \ \text{Prob}((\mathbf{x}, \boldsymbol{\omega}) \in \mathbf{K}) \geq 1 - \varepsilon \}. \qquad (7.13)$$

Indeed one then replaces problem (7.13) with

$$\min \{ f(\mathbf{x}) : \mathbf{x} \in C; \ h_d(\mathbf{x}) \geq 0 \}, \qquad (7.14)$$

where the uncertain parameter $\boldsymbol{\omega}$ has disappeared. So if C is a basic semi-algebraic set then (7.14) is a standard polynomial optimization problem. Of course the resulting decision problem (7.14) may still be hard to solve because the sets $\mathbf{X}_\varepsilon^*$ and $\mathbf{X}_\varepsilon^d$ are not convex in general. But this may be the price to pay for avoiding a too conservative formulation of the problem. However, in the formulation (7.14) one has gotten rid of the disturbance parameter $\boldsymbol{\omega}$, and so one may apply the arsenal of Non-Linear Programming algorithms to get a local minimizer of (7.14). If n is not too large or if some sparsity is present in problem (7.14) one may even run the Moment-SOS hierarchy to approximate its global optimal value. For the latter approach the interested reader is referred to Lasserre (2010) and for a discussion on this approach to various control problems with chance-constraints we refer to the recent paper of Jasour et al. (2015) and the references therein.

There is a rich literature on chance-constrained programming since Charnes and Cooper (1959) and Miller and Wagner (1965). Of particular interest are uncertainty models and methods that allow to define *tractable* approximations of $\mathbf{X}_\varepsilon^*$. Therefore an important issue is to analyze under which conditions on f, μ and the threshold ε, the resulting chance constraint defines a convex set $\mathbf{X}_\varepsilon^*$; see e.g. Henrion and Strugarek (2008), van Ackooij (2009), Nemirovski and Shapiro (2006), and van Ackooij and Malick (2018). For instance, in van Ackooij and Malick (2018) the authors consider joint chance-constraint $(f(\mathbf{x}, \boldsymbol{\omega}) \in \mathbb{R}^k)$ and show that $\mathbf{X}_\varepsilon^*$ is convex for sufficiently small ε if $\boldsymbol{\omega}$ is an elliptical random vector and f is convex in $\mathbf{x}$. The interested reader is referred to Henrion (2007), Prékopa (2003) and Shapiro et al. (2014) for a general overview of chance constraints in optimization and to Jasour and Lagoa (2017) in control (and the many references therein).

One well-sounded critic to these probabilistic approaches is that they rely on the knowledge of the exact distribution μ of the noise $\boldsymbol{\omega}$, which in many cases may not be a realistic assumption. Therefore modeling the uncertainty via a single *known* probability distribution μ is questionable, and may render a chance-constraint like $\text{Prob}_\mu(f(\mathbf{x}, \boldsymbol{\omega}) > 0) \geq 1 - \varepsilon$, counter-productive. It would rather make sense to assume some *partial knowledge* on the *unknown* distribution μ of the noise $\boldsymbol{\omega} \in \boldsymbol{\Omega}$. This is what has been investigated in for instance Calafiore and El Ghaoui (2006), Delage and Ye (2010) and Erdogan and Iyengar (2006). In Lasserre and Weisser (2019), the authors have extended the Moment-SOS hierarchy in order to handle *distributionally robust* chance-constraints of the form

$$\text{Prob}_\mu(f(\mathbf{x}, \boldsymbol{\omega}) > 0) > 1 - \varepsilon, \qquad \forall \mu \in \mathcal{M}_{\mathbf{a}},$$

where $\mathcal{M}_{\mathbf{a}}$ is the set of all possible mixtures of probability measures $\mu_{\mathbf{a}} \in \mathscr{P}(\boldsymbol{\Omega})$ parametrized by $\mathbf{a} \in \mathbf{A}$. For instance and typically, $\mathbf{a} = (a, \sigma)$ and $d\mu_{\mathbf{a}}(\boldsymbol{\omega}) = \exp(-(\boldsymbol{\omega} - a)^2/2\sigma^2)d\boldsymbol{\omega}$ (for mixtures of Gaussians). Crucial to extend the Moment-SOS methodology to this framework is the fact that for every α, the moment mapping $\mathbf{a} \mapsto \int_{\boldsymbol{\Omega}} \mathbf{x}^\alpha \, d\mu_{\mathbf{a}}(\boldsymbol{\omega})$ is a *polynomial* in the parameter $\mathbf{a} \in \mathbf{A}$. For more details on this non-trivial extension the interested reader is referred to Lasserre and Weisser (2019).

Bibliography

van Ackooij, W. (2009). *Convexity Statements for Linear Probability Constraints with Gaussian Technology Matrices and Copulae Correlated Rows*, DOI10.13140/RG.2.2.11723.69926.

van Ackooij, W., and Malick, J. (2018). *Eventual convexity of probability constraints with elliptical distributions*, Math. Program. pp. 1–27. doi.org/10.1007/s10107-018-1230-3.

Delage, E., and Ye, Y. (2010). *Distributionally robust optimization under moment uncertainty with applications to data-driven problems*, Oper. Res. **58**, pp. 595–612.

Calafiore, G., and Dabbene, F. (2006). *Probabilistic and Randomized Methods for Design under Uncertainty*, G. Calafiore and F. Dabbene (Eds.), Springer.

Calafiore, G. C., and El Ghaoui, L. (2006). *On Distributionally Robust Chance-Constrained Linear Programs*, J. Optim. Theory Appl. **130**, pp. 1–22.

Charnes, A., and Cooper, W. W. (1959). *Chance constrained programming*, Manag. Sci. **6**, pp. 73–79.

Erdogan, E., and Iyengar, G. (2006). *Ambiguous chance constrained problems and robust optimization*, Math. Program. Sér. B **107**, pp. 37–61.

Henrion, R. (2007). *Structural Properties of Linear Probabilistic Constraints*, Optimization **56**, pp. 425–440.

Henrion, R., and Strugarek, C. (2008). *Convexity of chance constraints with independent random variables*, Comput. Optim. Appl. **41**, pp. 263–276.

Jasour, A. M., Aybat, N. S., Lagoa, C. M. (2015). *Semidefinite programming for chance constrained optimization over semi-algebraic sets*, SIAM J. Optim. **25**, pp. 1411–1440.

Jasour, A., and Lagoa, C. (2017). *Convex constrained semialgebraic volume optimization: Application in systems and control*, arXiv:1701.08910.

Lasserre, J. B., and Weisser, T. (2019). *Distributionally robust polynomial chance-constraints under mixture ambiguity sets*, Math. Program. Ser. A, to appear. arXiv:1803.11500.

Lasserre, J. B. (2017a). *Representation of chance-constraints with strong asymptotic properties*, IEEE Control Systems Letters 1, pp. 50–55.

Lasserre, J. B. (2017b). *Computing gaussian and exponential measures of semi-algebraic sets*, Adv. Appl. Math. **91**, pp. 137–163.

Lasserre, J. B. (2010). *Moments, Positive Polynomials and Their Applications*, (Imperial College Press, London).

Li, P., Wendt, M., Wozny, G. (2002). *A Probabilistically Constrained Model Predictive Controller*, Automatica **38**, pp. 1171–1176.

Miller, B., and Wagner, H. (1965). *Chance-constrained programming with joint constraints*, Oper. Res. **13**, pp. 930–945.

Nemirovski, A., and Shapiro, A. (2006). *Convex approximations of chance constrained programs*, SIAM J. Optim. **17**, pp. 969–996.

Prékopa, A. (2003). *Probabilistic Programming*, in *Stochastic Programming*, A. Ruszczynski and A. Shapiro (Eds.), Handbooks in Operations Research and Management Science Volume 10, pp. 267–351.

Shapiro, A., Dentcheva, D., Ruszczynski, A. (2014). *Lecture on Stochastic Programming: Modeling and Theory* (2nd Ed., SIAM, Philadelphia).

Chapter 8

Approximate Optimal Design

We describe how to apply the Moment-SOS hierarchy to obtain approximate (polynomial) optimal design without any discretization of the design space.

8.1 Introduction

Optimal Design is concerned with minimizing the uncertainty contained in the best linear unbiased estimators in regression problems. The experimenter models the responses $z_1, \ldots, z_N$ of a random *experiment* whose inputs are represented by a vector $t_i \in \mathbb{R}^n$ with respect to known *regression functions* $\mathbf{f}_1, \ldots, \mathbf{f}_p$, namely:

$$z_i = \sum_{j=1}^{p} \theta_j \mathbf{f}_j(t_i) + \varepsilon_i, \quad i = 1, \ldots, N,$$

where $\theta_1, \ldots, \theta_p$ are unknown parameters that the experimenter wants to estimate, ε_i, $i = 1, \ldots, N$, are i.i.d. centered square integrable random variables and the inputs t_i are chosen by the experimenter in a *design space* $\mathbf{X} \subseteq \mathbb{R}^n$.

Assume that the inputs t_i, for $i = 1, \ldots, N$, are chosen within a set of distinct points $\{\mathbf{x}_1, \ldots, \mathbf{x}_\ell\} \subset \mathbf{X}$ with $\ell \leq N$, and let n_k denote the number of times the particular point $\mathbf{x}_k$ is chosen. This process is summarized by defining a design ξ as follows:

$$\xi := \begin{pmatrix} \mathbf{x}_1 & \cdots & \mathbf{x}_\ell \\ \frac{n_1}{N} & \cdots & \frac{n_\ell}{N} \end{pmatrix}, \tag{8.1}$$

whose first row gives distinct points in the design space $\mathbf{X}$ where the input parameters have to be taken and the second row indicates the experimenter which proportion of experiments (frequencies) have to be done at these points. The *information matrix* associated with ξ is defined by:

$$\mathbf{M}(\xi) := \sum_{i=1}^{\ell} w_i \, \mathbf{F}(\mathbf{x}_i) \, \mathbf{F}(\mathbf{x}_i)^T, \tag{8.2}$$

where $\mathbf{F} := (\mathbf{f}_1, \ldots, \mathbf{f}_p)$ is the column vector of regression functions and $w_i := n_i/N$ is the weight corresponding to the point x_i. One may and will identify a design ξ as in (8.1) with a discrete probability measure on $\mathbf{X}$ with finite support given by the points $\mathbf{x}_i$ and weights w_i.

Notice that $\mathbf{M}(\xi) \succeq 0$. For all $q \in [-\infty, 1]$ define the function $\mathbf{M} \mapsto \phi_q(\mathbf{M})$, $\mathbf{M} \succ 0$, by:

$$\phi_q(\mathbf{M}) := \begin{cases} (\frac{1}{p}\mathrm{trace}(\mathbf{M}^q))^{1/q} & \text{if } q \neq -\infty, 0 \\ \det(\mathbf{M})^{1/p} & \text{if } q = 0 \\ \lambda_{\min}(\mathbf{M}) & \text{if } q = -\infty \end{cases}$$

and for $\mathbf{M} \succeq 0$

$$\phi_q(\mathbf{M}) := \begin{cases} (\frac{1}{p}\mathrm{trace}(\mathbf{M}^q))^{1/q} & \text{if } q \in (0, 1] \\ 0 & \text{if } q \in [-\infty, 0], \end{cases}$$

where $\mathrm{trace}(\mathbf{M})$, $\det(\mathbf{M})$ and $\lambda_{\min}(\mathbf{M})$ denote respectively the trace, determinant and least eigenvalue of $\mathbf{M}$. All these criteria are real valued, positively homogeneous, non-constant, upper semi-continuous, isotonic (with respect to the Loewner ordering) and concave functions.

Hence, an *optimal design* is a solution ξ^* to the following problem

$$\max \phi_q(\mathbf{M}(\xi)) \tag{8.3}$$

where the maximum is taken over all ξ of the form (8.1). Standard criteria are given by the parameters $q = 0, -1, -\infty$ and are referred to D, A or E-optimum designs respectively.

In this chapter we consider the case where the regression functions $\mathbf{F}$ are all multivariate monomials of degree at most d, and

$$\mathbf{X} := \{\mathbf{x} \in \mathbb{R}^n : g_j(\mathbf{x}) \geq 0, \quad j = 1, \ldots, m\} \tag{8.4}$$

for some polynomials $(g_j) \subset \mathbb{R}[\mathbf{x}]$.

• We restrict our attention to *approximate optimal designs* where, by definition, we replace the set of *feasible* matrices $\mathbf{M}(\xi)$ with ξ of the form (8.1) by the larger set of all possible information matrices, namely the convex hull of $\{\mathbf{F}(\mathbf{x})\,\mathbf{F}(\mathbf{x})^T : \mathbf{x} \in \mathbf{X}\}$.

• To construct approximate optimal designs, we provide a two-step procedure presented in Algorithm 8.1 below. This procedure first finds the information matrix $\mathbf{M}(\xi^*)$ of an approximate optimal design ξ^* and next computes the support points $\mathbf{x}_i^*$ and the weights w_i^* of the design ξ^* in a second step.

> This two-step procedure has two distinguishing features. (i) It avoids any discretization of the design space $\mathbf{X}$, and (ii) the design space $\mathbf{X}$ can be any compact basic semi-algebraic set, hence not necessarily convex nor connected. Briefly, an approximate design ξ is seen as a *probability measure* on $\mathbf{X}$ and maximizing the concave criterion $\phi_q(\mathbf{M}(\xi))$ (on the moment matrix of ξ) yields an optimal approximate design ξ^* with *finite* support on $\mathbf{X}$.

8.2 The ideal approximate optimal design problem

Recall that $\mathcal{M}(\mathbf{X})_{2d}$ is the convex cone of sequence $\mathbf{y} = (y_\alpha) \in \mathbb{R}^{s(2d)}$ that have a representing measure supported on $\mathbf{X}$. Its dual cone is the convex cone $\mathscr{P}(\mathbf{X})_{2d}$ of polynomials of degree at most $2d$, nonnegative on $\mathbf{X}$. See Lemma 1.7. Let $\mathbf{x} \mapsto \mathbf{v}_d(\mathbf{x}) \in \mathbb{R}^{s(d)}$ denote the vector of all monomials up to degree d.

> The *ideal* approximate optimal design problem reads:
> $$\rho = \max_{\mathbf{y}} \{\, \phi_q(\mathbf{M}_d(\mathbf{y})) : \mathbf{y} \in \mathcal{M}(\mathbf{X})_{2d}; \ y_0 = 1 \,\}. \qquad (8.5)$$

Problem (8.5) is a convex optimization problem (however the objective function is not linear in the variable $\mathbf{y}$). The following result is standard.

Theorem 8.1 (Equivalence theorem). *Let $q \in (-\infty, 1)$ and $\mathbf{X} \subseteq \mathbb{R}^n$ be the compact semi-algebraic defined in (8.4) and with nonempty interior.*

(a) Problem (8.5) is a convex optimization problem with a unique optimal solution $\mathbf{y}^* \in \mathcal{M}(\mathbf{X})_{2d}$. Denote by p_d^* the SOS polynomial

$$\mathbf{x} \mapsto p_d^*(\mathbf{x}) := \mathbf{v}_d(\mathbf{x})^\top \mathbf{M}_d(\mathbf{y}^*)^{q-1} \mathbf{v}_d(\mathbf{x}) = \|\mathbf{M}_d(\mathbf{y}^*)^{\frac{q-1}{2}} \mathbf{v}_d(\mathbf{x})\|_2^2. \quad (8.6)$$

Then $\mathbf{y}^*$ is the vector of moments — up to order $2d$ — of a discrete measure μ^* supported on at least $\binom{n+d}{n}$ and at most $\binom{n+2d}{n}$ points in the set

$$\Omega := \left\{ \mathbf{x} \in \mathbf{X} : p^*(\mathbf{x}) = 0 \right\} = \left\{ \mathbf{x} \in \mathbf{X} : p_d^*(\mathbf{x}) = \mathrm{trace}(\mathbf{M}_d(\mathbf{y}^*)^q) \right\}$$

with $\mathbf{x} \mapsto p^*(\mathbf{x}) := \mathrm{trace}(\mathbf{M}_d(\mathbf{y}^*)^q) - p_d^*(\mathbf{x}) \in \mathcal{P}(\mathbf{X})_{2d}$.

(b) The following statements are equivalent:

- $\mathbf{y}^* \in \mathcal{M}(\mathbf{X})_{2d}$ *is the unique solution to Problem (8.5);*
- $\mathbf{y}^* \in \{\mathbf{y} \in \mathcal{M}(\mathbf{X})_{2d} : y_0 = 1\}$ *and $p^* := \mathrm{trace}(\mathbf{M}_d(\mathbf{y}^*)^q) - p_d^* \geqslant 0$ on $\mathbf{X}$.*

On the optimal dual polynomial. The polynomial p_d^* in (8.6) contains all the information concerning the optimal design. Indeed, its level set Ω contains the support of the optimal approximate design ξ^* and $\mathbf{M}_d(\mathbf{y}^*)$ (the inverse of its Gram matrix) encodes the support of ξ^* and the weights $\mathbf{w}$. When $q = 0$ the reciprocal of p_d^* is the so-called Christoffel function, well-known in theory of approximation; see §1.2.4. For this reason, in the sequel p_d^* in (8.6) is called a Christoffel polynomial.

8.3 The Moment-SOS hierarchy for optimal design

We next describe how to apply the Moment-SOS hierarchy so solve the ideal optimal approximate design problem. As the design space $\mathbf{X}$ in (8.4) is compact it is contained in the ball $\{\mathbf{x} : \|\mathbf{x}\|^2 \leq M\}$ for some M (assumed to be known) and therefore we may and will assume that $g_1(\mathbf{x}) = M - \|\mathbf{x}\|^2$. This implies that $Q(g)$ is Archimedean; see Definition 1.3.

Recall that $s(t) := \binom{n+t}{n}$, $t \in \mathbb{N}$, and let $d_j := \lceil \deg(g_j)/2 \rceil$, $j = 1, \ldots, m$. With d fixed, and for every $\delta = 0, 1, \ldots$, define:

$$\mathcal{M}(\mathbf{X})_{2d}^{SDP,\delta} := \{ \, \mathbf{y}_{d,\delta} \in \mathbb{R}^{s(2d)} : \exists \, \mathbf{y}_\delta \in \mathbb{R}^{s(2(d+\delta))} \text{ s.t.} \atop \mathbf{M}_{d+\delta}(\mathbf{y}_\delta), \, \mathbf{M}_{d-\delta-d_j}(\mathbf{y}_\delta) \succeq 0, \, j = 1, \ldots, m \, \}. \tag{8.7}$$

Observe that $\mathcal{M}(\mathbf{X})_{2d}^{SDP,(\delta+1)} \subset \mathcal{M}(\mathbf{X})_{2d}^{SDP,\delta}$ and $\mathcal{M}(\mathbf{X})_{2d} \subset \mathcal{M}(\mathbf{X})_{2d}^{SDP,\delta}$ for all $\delta = 0, 1, \ldots$ In fact as $Q(g)$ is Archimedean, by Putinar's Theorem 1.9

$$\mathcal{M}(\mathbf{X})_{2d} = \overline{\bigcap_{\delta \geq 0} \mathcal{M}(\mathbf{X})_{2d}^{SDP,\delta}}.$$

To solve the approximate optimal design problem we propose a two-step procedure. At each of the two steps one applies the Moment-SOS hierarchy in an appropriate manner.

8.3.1 The first step

The first step: With δ fixed, consider the following optimization problem:

$$\rho_\delta = \max_{\mathbf{y}} \{ \, \phi_q(\mathbf{M}_d(\mathbf{y})) : y_0 = 1; \, \mathbf{y} \in \mathcal{M}(\mathbf{X})_{2d}^{SDP,\delta} \, \}. \tag{8.8}$$

From what precedes (8.8) is an obvious relaxation of (8.5) and therefore $\rho_\delta \geq \rho$ for all $\delta = 0, 1, \ldots$

Theorem 8.2 below is for (8.8) what Theorem 8.1 is for (8.5).

Theorem 8.2 (Equivalence theorem for relaxation). *Let $q \in (-\infty, 1)$ and $\mathbf{X} \subseteq \mathbb{R}^n$ be the compact semi-algebraic defined in (8.4) and with nonempty interior. Then:*

(a) Problem (8.8) is a convex optimization problem with a unique optimal solution $\mathbf{y}^* \in \mathcal{M}(\mathbf{X})_{2d}^{SDP,\delta}$.

(b) The moment matrix $\mathbf{M}_d(\mathbf{y}^*)$ is positive definite. Let p_d^* be the SOS polynomial associated with $\mathbf{y}^*$ as defined in (8.6). Then $p^* := \text{trace}(\mathbf{M}_d(\mathbf{y})^q) - p_d^* \in \mathcal{P}(\mathbf{X})_{2d}$ and $L_{\mathbf{y}^*}(p^*) = 0$. In particular, the following statements are equivalent:

- *$\mathbf{y}^* \in \mathcal{M}(\mathbf{X})_{2d}^{SDP,\delta}$ is the unique solution to Problem (8.8);*
- *$\mathbf{y}^* \in \{\mathbf{y} \in \mathcal{M}(\mathbf{X})_{2d}^{SDP,\delta} : y_0 = 1\}$ and $p^* := \text{trace}(\mathbf{M}_d(\mathbf{y}^*)^q) - p_d^* \in Q_{2(d+\delta)}(g) \cap \mathbb{R}[\mathbf{x}]_{2d}$.*

(c) If $\mathbf{y}^* \in \mathcal{M}(\mathbf{X})_{2d}$ then $\mathbf{y}^*$ is also an optimal solution of (8.5).

Proof. Brief sketch of the proof. (a) Notice that the feasible set $\mathcal{M}_{2d}^{SDP,\delta}$ is the projection of $\mathcal{M}(\mathbf{X})_{2(d+\delta)}^{SDP}$ on $\mathbb{R}^{s(2d)}$. By Lemma 1.3 the latter is compact and therefore so is its projection $\mathcal{M}_{2d}^{SDP,\delta}$. As the objective function is strictly concave, uniqueness of the optimal solution follows. (b) The optimal solution p^* is constructed from the KKT-optimality conditions and by definition is en element of $Q_{2(d+\delta)}(g) = (\mathcal{M}(\mathbf{X})_{2(d+\delta)}^{SDP})^*$ (where $Q_d(g)$ is defined in (1.5)). Next, $L_{\mathbf{y}^*}(p^*) = 0$ is just the complementarity condition. The last statement follows because Slater's condition holds for (8.8) and therefore the necessary KKT-optimality conditions are also sufficient. (c) If $\mathbf{y}^* \in \mathcal{M}(\mathbf{X})_{2d}$ then by Theorem 8.1(b) $\mathbf{y}^*$ solves the optimal approximate design problem. $\qquad\qquad\square$

One expects that the unique optimal solution $\mathbf{y}^*$ of (8.8) is also that of (8.5) (and this is indeed the case in all our numerical experiments on nontrivial examples, even with $\delta = 0$). To check whether $\mathbf{y}^* \in \mathcal{M}(\mathbf{X})_{2d}$ one may check whether the sufficient flat extension condition (1.12) holds at $\mathbf{y}_\delta^*$ associated with $\mathbf{y}^*$ in the definition (8.7) of $\mathcal{M}_{2d}^{SDP,\delta}$. However even if $\mathbf{y}^* \in \mathcal{M}(\mathbf{X})_{2d}$, in general (1.12) will not hold at $\mathbf{y}_\delta^*$ because in maximizing the criterion ϕ_q in (8.8), the $2(d+\delta)$ moments $y_{\delta,\alpha}^*$ with $|\alpha| = 2(d+\delta)$ are not penalized.

Fortunately there is also another test to detect whether indeed $\mathbf{y}^* \in \mathcal{M}(\mathbf{X})_{2d}$. This is the purpose of the second step of the procedure described below.

8.3.2 The second step

The input of the second step is the unique optimal solution $\mathbf{y}^* \in \mathcal{M}(\mathbf{X})_{2d}^{SDP,\delta}$, of (8.8) for some δ (the output of the first step of the procedure). Its goal is to certify that in fact $\mathbf{y}^* \in \mathcal{M}(\mathbf{X})_{2d}$ and so $\mathbf{y}^*$ solves the approximate primal design problem. It uses a specific Moment-SOS hierarchy due to Nie (2013).

The second step: Let $\mathbf{y}^* \in \mathcal{M}(\mathbf{X})_{2d}^{SDP,\delta}$ be the unique optimal solution of (8.8) for some δ, and with $s \in \mathbb{N}$ fixed, consider the following optimization problem:

$$\rho'_t = \min_{\mathbf{z}} \left\{ L_{\mathbf{z}}(f_{d+t}) : z_\alpha = y_\alpha^*, \quad \forall \alpha \in \mathbb{N}_{2d}^n \right.$$
$$\left. \mathbf{M}_{d+t}(\mathbf{z}), \mathbf{M}_{d+t-d_j}(g_j\,\mathbf{z}) \succeq 0, \quad j = 1,\ldots,m \right\} \tag{8.9}$$

where $f_{d+t} \in \mathbb{R}[\mathbf{x}]_{2(d+t)}$ is a (randomly generated) polynomial of degree $2(d+t)$, strictly positive on $\mathbf{X}$.

As for (8.8), Problem (8.9) has an optimal solution $\mathbf{z}^* \in \mathbb{R}^{s(2(d+t))}$ because its feasible set is compact. Let $d^* := \max[d_1,\ldots,d_m]$.

Lemma 8.3. *Let $\mathbf{z}_t^* \in \mathbb{R}^{s(2(d+t))}$ be an optimal solution of (8.9). If*

$$\mathrm{rank}(\mathbf{M}_{d+\ell}(\mathbf{z}_t^*)) = \mathrm{rank}(\mathbf{M}_{d+\ell-d^*}(\mathbf{z}^*)) \tag{8.10}$$

for some $\ell \leq t$ then $\mathbf{y}^* \in \mathcal{M}(\mathbf{X})_{2d}$ and $\mathbf{y}^*$ solves the approximate optimal design. The atoms of the optimal design ξ can be extracted by a linear algebra routine.

Conversely, if $\mathbf{y}^* \in \mathcal{M}(\mathbf{X})_{2d}$ then generically there exists t such that (8.10) holds at an optimal solution $\mathbf{z}_t^* \in \mathbb{R}^{s(2(d+t))}$ of (8.9).

Proof. By Theorem 1.6, $\mathbf{z}^* \in \mathcal{M}(\mathbf{X})_{2(d+t)}$ and so $\mathbf{y}^* \in \mathcal{M}(\mathbf{X})_{2d}$. Then the result follows from Theorem 8.2(c). The converse result is a consequence of a result from Nie (2013).   □

Recovering ξ. Let $\kappa := \mathrm{rank}(\mathbf{M}_{d+\ell}(\mathbf{z}_t^*))$ in (8.10). The κ atoms $(\mathbf{x}_j)_{j=1,\ldots,\kappa} \subset \mathbf{X}$ of ξ can be recovered by a linear algebra routine as described in Henrion and Lasserre (2005) and Henrion et al. (2009). Then the weights $\mathbf{w}$ are obtained by solving the linear system:

$$\sum_{j=1}^{\kappa} w_j \, \mathbf{x}_j^{\alpha} \; = \; y_{\alpha}^{*}, \quad \alpha \in \mathbb{N}_{2d}^{n}.$$

So in solving (8.9) one also considers extensions $\mathbf{z} \in \mathbb{R}^{s(2(d+t))}$ of the sequence $\mathbf{y}^{*} \in \mathbb{R}^{s(2d)}$, exactly as $\mathbf{y}_{\delta}^{*} \in \mathbb{R}^{s(2(d+\delta))}$ at an optimal solution of (8.8). However, in contrast to (8.8), the criterion in (8.9) penalizes the high order moments z_{α} with $|\alpha| = 2(d+t)$ and enforces the rank condition (8.10) at an optimal solution $\mathbf{z}_{t}^{*}$. The 2-step Algorithm can be summarized as follows:

Algorithm 8.1 (Approximate Optimal Design).

Input: A compact semi-algebraic design space $\mathbf{X}$ defined as in (8.4).
Output: An approximate optimal design ξ.

(1) Choose the two relaxation orders δ and t.
(2) Solve the convex relaxation (8.8) of order δ and get $\mathbf{y}_{\delta}^{*} \in \mathbb{R}^{s(2(d+\delta))}$.
(3) Solve the semidefinite program (8.9) and get $\mathbf{z}_{t}^{*} \in \mathbb{R}^{s(2(d+t))}$.
(4) If $\mathbf{z}_{t}^{*}$ satisfies the rank condition (8.10), then extract the optimal design ξ from the truncated moment sequence $\mathbf{z}_{t}^{*}$ (e.g. as described in Henrion and Lasserre (2005)).
(5) Otherwise, choose larger values of δ and t and go to Step 2.

8.4 Numerical experiments

We have run Algorithm 8.1 on two $2D$-examples and one $3D$-examples with the criterion ϕ_q with $q = 0$ (the D-optimal design problem).

Example 8.1 (Wynn's polygon). Consider the polygon with vertices $(-1,-1)$, $(-1,1)$, $(1,-1)$ and $(2,2)$, scaled to fit the unit circle, *i.e.*,

$$\mathbf{X} = \{ \mathbf{x} \in \mathbb{R}^2 : x_1, x_2 \geqslant -\tfrac{1}{4}\sqrt{2}, \; x_1 \leq \tfrac{1}{3}(x_2 + \sqrt{2})$$
$$x_2 \leq \tfrac{1}{3}(x_1 + \sqrt{2}), \; x_1^2 + x_2^2 \leq 1 \}.$$

Note that have introduced the redundant constraint $x_1^2 + x_2^2 \leq 1$ in order to have an algebraic certificate of compactness.

We solve Problems (8.8) and (8.9). Let us start with $d = 1$ and $\delta = 3$. Solving (8.8) we obtain $\mathbf{y}^* \in \mathbb{R}^{45}$ which leads to 4 atoms when solving (8.9) with $t = 3$. For the latter the moment matrices $\mathbf{M}_2(\mathbf{z}_3^*)$ and $\mathbf{M}_3(\mathbf{z}_3^*)$ of order 2 and 3 both have rank 4, so Condition (8.10) is fulfilled. As expected, the 4 atoms are exactly the vertices of the polygon.

If one increases d, we get an atomic optimal measure with a larger support. For $d = 2$ we recover 7 points, and 13 for $d = 3$. See Figure 8.1 for the polygon, the supporting points of the optimal measure and the $\binom{2+d}{2}$-level set of the Christoffel polynomial p_d^* for different d. The latter demonstrates graphically that the set of zeros of $\binom{2+d}{d} - p_d^*$ intersected with $\mathbf{X}$ are indeed the atoms of our representing measure. (In the picture the size of the support points is chosen with respect to their corresponding weights, i.e., the larger the point, the bigger the respective weight.)

To get an idea of how the Christoffel polynomial looks like, we plot in Figure 8.2 the 3D-plot of the polynomial $-p^* = p_d^* - \binom{2+d}{2}$. This illustrates very clearly that the zeros of p^* on $\mathbf{X}$ are the support points of the optimal design.

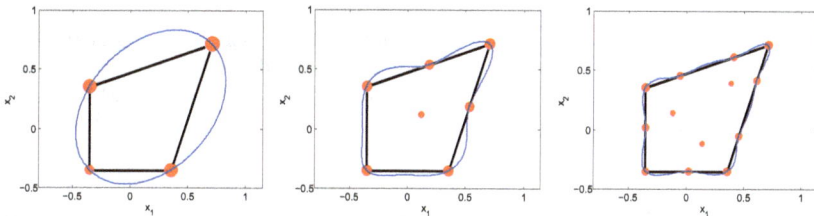

Fig. 8.1   The polygon (bold black) of Example 8.1, the support of the optimal design measure (red points) where the size of the points corresponds to the respective weights, and the $\binom{2+d}{2}$-level set of the Christoffel polynomial (thin blue) for $d = 1$ (left), $d = 2$ (middle), $d = 3$ (right) and $\delta = 3$.

Example 8.2 (Folium). The zero level set of the polynomial $f(\mathbf{x}) = -x_1(x_1^2 - 2x_2^2)(x_1^2 + x_2^2)^2$ is a curve of genus zero with a triple singular

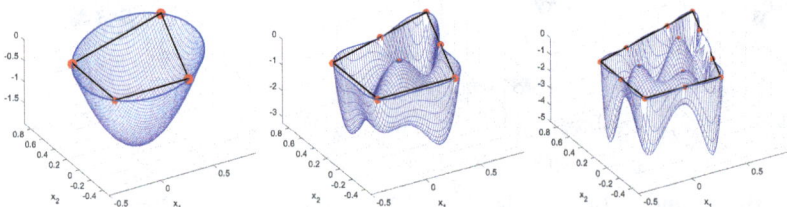

Fig. 8.2 The polynomial $p_d^* - \binom{2+d}{2}$ where p_d^* denotes the Christoffel polynomial of Example 8.1 for $d = 1$ (top left), $d = 2$ (top right), $d = 3$ (bottom middle). The red points correspond to the $\binom{2+d}{2}$-level set of the Christoffel polynomial.

point at the origin. It is called a *folium*. Consider the semi-algebraic set

$$\mathbf{X} = \{\mathbf{x} \in \mathbb{R}^2 : f(\mathbf{x}) \geq 0,\ x_1^2 + x_2^2 \leq 1\}.$$

Figure 8.3 illustrates the results.

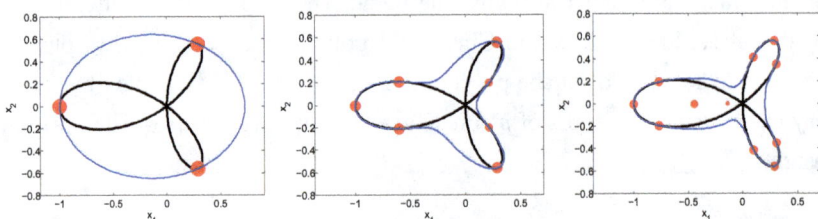

Fig. 8.3 The polygon (bold black) of Example 8.2, the support of the optimal design measure (red points) where the size of the points corresponds to the respective weights, and the $\binom{2+d}{2}$-level set of the Christoffel polynomial (thin blue) for $d = 1$ (left), $d = 2$ (middle), $d = 3$ (right) and $\delta = 3$.

Example 8.3 (The 3-dimensional unit sphere). Consider the regression for the degree d polynomial measurements $\sum_{|\alpha| \leq d} \theta_\alpha \mathbf{x}^\alpha$ on the unit sphere $\mathbf{X} = \{\mathbf{x} \in \mathbb{R}^3 : \|\mathbf{x}\|^2 = 1\}$. For $d = 1$ and $\delta \geq 0$ we obtain the sequence $\mathbf{y}^* \in \mathbb{R}^{10}$ with $y_{000}^* = 1$, $y_{200}^* = y_{020}^* = y_{002}^* = 0.333$ and all other entries zero. In the second step we solve (8.9) to recover the measure. For $t = 2$ the moment matrices of order $d + t = 2$ and $d + t - 1 = 3$ both have rank 6, meaning the rank condition (8.10) is fulfilled, and we obtain the six atoms $\{(\pm 1, 0, 0), (0, \pm 1, 0), (0, 0, \pm 1)\} \subseteq \mathbf{X}$ on which the optimal measure is uniformly supported.

For quadratic regressions, *i.e.*, $d = 2$, we obtain an optimal measure supported on 14 atoms evenly distributed on the sphere. Choosing $d = 3$ (cubic regressions), we obtain an atomic measure supported on 26 atoms evenly distributed on the sphere. Results for $d = 1$, $d = 2$, $d = 3$ and $\delta = 0$ are displayed in Figure 8.4.

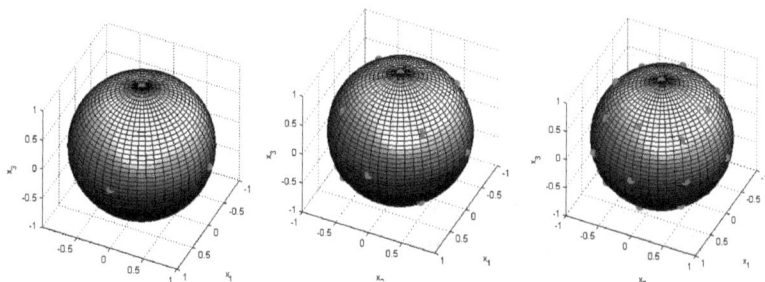

Fig. 8.4   The red points illustrate the support of the optimal design measure for $d = 1$ (left), $d = 2$ (middle), $d = 3$ (right) and $\delta = 0$ for Example 8.3.

8.5   Concluding remarks

We have provided a two-step procedure to compute an approximate optimal design ξ^* on a compact basic semi-algebraic set $\mathbf{X}$. In contrast to other numerical approaches as e.g. in Torsney (2009) and Sagnol and Harman (2015), a distinguishing feature of the approach is to not rely on an *a priori* discretization of the design space $\mathbf{X}$. The approximate optimal design ξ^* is viewed as an atomic probability measure on $\mathbf{X}$ that maximizes a concave Kiefer-criterion $\phi_q(\mathbf{M}(\mu))$ over all probability measures μ on $\mathbf{X}$ (where $\mathbf{M}(\mu)$ is the moment matrix up to order d of μ). The optimality conditions reveal that the atomic support of ξ^* is the zero-set of a certain dual polynomial $s(d) - p_d^*$, positive on $\mathbf{X}$, where p_d^* is the (SOS) Christoffel polynomial of degree $2d$ associated with ξ^*.

The two-step procedure is relatively simple and illustrates the versatility of the Moment-SOS hierarchy. Interestingly, and in contrast to other chapters, we apply the Moment-SOS hierarchy *twice*, i.e., at each of the two steps. It consists of a hierarchy of *convex* (but not SDP) relaxations at the first step and a hierarchy of SDP-relaxations at the second step. On a practical side and for numerical stability, it would certainly be wiser to express the moment and localizing matrices in a basis of polynomials that is more convenient than the monomial basis.

8.6   Notes and sources

This chapter is essentially from de Castro et al. (2018). For a complete overview on the subject of the theory of optimal design of experiments the interested reader is referred to Kiefer (1974), the inspiring book of Dette and Studden (1997), and the many references therein. In de Castro et al. (2018) the authors also propose an alternative strategy in step 2 to recover the optimal measure. Instead of running Nie's SDP-hierarchy (8.9) one may also minimize the polynomial $p^* := s(d) - p_d^*$ on $\mathbf{\Omega}$, where p_d^* is the Christoffel polynomial associated with the optimal solution $\mathbf{y}^*$ of (8.8) at step 1. Indeed if $\mathbf{y}^*$ has a representing measure μ on $\mathbf{X}$ then p^* vanishes on $\mathbf{X}$ at the support points of μ. For more details see de Castro et al. (2018).

Bibliography

De Castro, Y., Gamboa, F., Henrion, D., Hess, R., Lasserre, J. B. (2019). *Approximate optimal design for multivariate polynomial regression,* Annals of Statistics **47**, pp. 127–155.

Dette, H., and Studden, W. J. (1997). *The Theory of Canonical Moments with Applications in Statistics, Probability, and Analysis,* (John Wiley & Sons).

Henrion, D., and Lasserre J. B. (2005). *Detecting Global Optimality and Extracting Solutions in GloptiPoly,* in *Positive Polynomials in Control,* D. Henrion and A. Garulli (Eds.), Lecture Notes in Control and Information Sciences, Springer-Verlag, Berlin 2005), pp. 293–300.

Henrion D., Lasserre, J. B., Lofberg J. (2009). *Gloptipoly 3: moments, optimization and semidefinite programming,* Optim. Methods and Softwares **24**, pp. 761–779.

Kiefer, J. (1974). *General equivalence theory for optimum designs approximate theory),* Annals of Statistics **2**, pp. 849–879.

Nie, J. (2013). *Certifying convergence of Lasserre's hierarchy via flat truncation,* Math. Program. Ser. A **142**, pp. 485–510.

Sagnol, G., and Harman, R. (2015). *Computing exact D-optimal designs by mixed integer second-order cone programming,* Annals of Statistics **43**, pp. 2198–2224.

Torsney, B. (2009). *W-iterations and ripples therefrom,* in *Optimal Design and Related Areas in Optimization and Statistics,* Luc Pronzato and Anatoly Zhigljavsky (Eds.), Springer, pp. 1–12.

Part II

The Moment-SOS Hierarchy for Applications in Control, Optimal Control and Non-Linear Partial Differential Equations

Chapter 9

Optimal Control

We describe how to apply the Moment-SOS hierarchy in Optimal Control for controlled dynamical systems governed by ordinary differential equations (with polynomial dynamics).

9.1 Introduction

The Moment-SOS hierarchy described in Lasserre (2000, 2001, 2010, 2015) was primarily designed for solving the Generalized Moment Problem (GMP), and more particularly for solving non-convex polynomial optimization problems which are the simplest instances of the GMP. It was later applied to polynomial optimal control in Lasserre et al. (2008) and also in Gaitsgory and Quincampoix (2009) by following the same key idea. Namely one reformulates the initial non-convex polynomial optimal control problem with compact constraint set into an LP on the (infinite-dimensional) space of occupation measures supported on this set. However, while in polynomial optimization the initial problem is finite-dimensional, in polynomial optimal control the initial problem is already infinite-dimensional, which induces additional difficulties.

The objective of this chapter is to revisit the approach of Lasserre et al. (2008) and to survey the use of occupation measures to linearize polynomial optimal control problems. This is an opportunity to describe duality in infinite-dimensional conic problems, as well as various approximation results on the value function of the optimal control problem. The primal LP consists of finding occupation measures supported on optimal relaxed controlled trajectories, whereas the dual LP consists of finding the largest lower bound on the value function of the optimal control problem. The value function is the solution (in a suitably defined weak sense) of a non-linear partial differential equation called the Hamilton-Jacobi-Bellman equation, see e.g. (Trélat, 2005, Chapters 8 and 9) and (Clarke, 2013, Chapters 19 and 24). It is traditionally used for verification of optimality, and for explicit computation of optimal control laws, but we do not describe these applications here.

The occupation measures introduced in this chapter will be useful for solving other problems involving differential equations, e.g. in Chapter 10.

9.2 Polynomial optimal control

We consider polynomial optimal control problems (POCPs) of the form:

$$
\begin{aligned}
v^*(t_0, \mathbf{x}_0) := \inf_{\mathbf{u}} \int_{t_0}^{T} &\, l(\mathbf{x}(t), \mathbf{u}(t))dt + l_T(\mathbf{x}(T)) \\
\text{s.t. } &\dot{\mathbf{x}}(t) = f(\mathbf{x}(t), \mathbf{u}(t)), \ x(t_0) = \mathbf{x}_0 \\
&\mathbf{x}(t) \in \mathbf{X}, \ t \in [t_0, T] \\
&\mathbf{u}(t) \in \mathbf{U}, \ t \in [t_0, T] \\
&\mathbf{x}(T) \in \mathbf{X}_T
\end{aligned}
\tag{9.1}
$$

where the dot denotes time derivative, $l \in \mathbb{R}[\mathbf{x}, \mathbf{u}]$ is a given Lagrangian (integral cost), $l_T \in \mathbb{R}[\mathbf{x}]$ is a given terminal cost, $f \in \mathbb{R}[x, u]^n$ is a given dynamics (vector field), $\mathbf{X} \subset \mathbb{R}^n$ is a given compact state constraint set, $\mathbf{U} \subset \mathbb{R}^m$ is a given compact control constraint set, $\mathbf{X}_T \subset \mathbf{X}$ is a given compact terminal state constraint set. Also given are the terminal time $T \geq 0$, the initial time $t_0 \in [0, T]$ and the initial condition $\mathbf{x}_0 \in \mathbf{X}$. In POCP (9.1), the minimum is with respect to all control laws $\mathbf{u} \in \mathscr{L}^\infty([t_0, T]; \mathbf{U})$ which are bounded functions of time t with values in $\mathbf{U}$, and the resulting state trajectories $\mathbf{x} \in \mathscr{L}^\infty([t_0, T]; \mathbf{X})$ which are bounded functions of time t with values in $\mathbf{X}$.

Let $\mathscr{A} \subset [0, T] \times \mathbf{X}$ denote the set of values $(t_0, \mathbf{x}_0)$ for which there is a controlled trajectory $(\mathbf{x}, \mathbf{u}) \in \mathscr{L}^\infty([t_0, T]; \mathbf{X} \times \mathbf{U})$ starting at $\mathbf{x}(t_0) = \mathbf{x}_0$ and admissible for POCP (9.1). The function $(t_0, \mathbf{x}_0) \mapsto v^*(t_0, \mathbf{x}_0)$ defined in (9.1) is called the value function, and its domain is $\mathscr{A}$.

9.3 LP formulation

As explained in the introduction, to derive an LP formulation of POCP (9.1) we have to introduce measures on trajectories, the so-called occupation measures. The first step is to replace classical controls with probability measures, and for this we have to define additional notations.

9.3.1 *Relaxed controls*

In POCP (9.1), given $(t_0, \mathbf{x}_0) \in \mathscr{A}$, let $(\mathbf{x}_k, \mathbf{u}_k)_{k \in \mathbb{N}} \in \mathscr{L}^\infty([t_0, T]; \mathbf{X} \times \mathbf{U})$ denote a minimizing sequence of admissible controlled trajectories, i.e. it holds:

$$\dot{\mathbf{x}}_k(t) = f(\mathbf{x}_k(t), \mathbf{u}_k(t)), \quad \mathbf{x}(t_0) = \mathbf{x}_0$$

and

$$\lim_{k \to \infty} \int_{t_0}^T l(\mathbf{x}_k(t), \mathbf{u}_k(t))dt + l_T(\mathbf{x}_k(T)) = v^*(t_0, \mathbf{x}_0).$$

In general the infimum in POCP (9.1) is not attained, so our next step is to assume that, at each time $t \in [t_0, T]$, the control is not a vector $\mathbf{u}(t) \in \mathbf{U}$, but a time-dependent probability measure $w(du \mid t) \in \mathscr{P}(\mathbf{U})$ which rules the distribution of the control in $\mathbf{U}$. We use the notation $w_t := w(. \mid t)$ to emphasize the dependence on time. This is called a relaxed control, or stochastic control, or Young measure in the functional analysis literature. POCP (9.1) is then relaxed to:

$$v_R^*(t_0, \mathbf{x}_0) := \min_{\omega} \int_{t_0}^{T} \langle l(\mathbf{x}(t), .), \omega_t \rangle dt + l_T(\mathbf{x}(T))$$

$$\text{s.t. } \dot{\mathbf{x}}(t) = \langle f(\mathbf{x}(t), .), \omega_t \rangle, \quad \mathbf{x}(t_0) = \mathbf{x}_0$$

$$\mathbf{x}(t) \in \mathbf{X}, \ t \in [t_0, T]$$

$$\omega_t \in \mathscr{P}(\mathbf{U}), \ t \in [t_0, T] \tag{9.2}$$

$$\mathbf{x}(T) \in \mathbf{X}_T$$

where the minimization is w.r.t. a relaxed control. Note that we replaced the infimum in POCP (9.1) with a minimum in relaxed POCP (9.2). Indeed, it can be proved that this minimum is always attained using (weak-star) compactness of the space of probability measures with compact support.

Since classical controls $\mathbf{u} \in \mathscr{L}^{\infty}([t_0, T]; \mathbf{U})$ are a particular case of relaxed controls $\omega_t \in \mathscr{P}(\mathbf{U})$ corresponding to the choice $\omega_t = \delta_{\mathbf{u}(t)}$ for a.e. $t \in [t_0, T]$, the minimum in relaxed POCP (9.2) is smaller than the infimum in classical POCP (9.1), i.e.

$$v^*(t_0, \mathbf{x}_0) \geq v_R^*(t_0, \mathbf{x}_0).$$

Contrived optimal control problems (e.g. with overly stringent state constraints) can be cooked up such that the inequality is strict, i.e. $v^*(t_0, \mathbf{x}_0) > v_R^*(t_0, \mathbf{x}_0)$, see e.g. the examples in (Henrion and Korda, 2014, Appendix C). We do not consider that these examples are practically relevant, and therefore throughout the whole chapter we make the following assumption:

Assumption 9.1 (No relaxation gap). *For any relaxed controlled trajectory* (x, ω_t) *admissible for relaxed POCP (9.2), there is a sequence of controlled trajectories* $(\mathbf{x}_k, \mathbf{u}_k)_{k \in \mathbb{N}}$ *admissible for POCP (9.1) such that*

$$\lim_{k \to \infty} \int_{t_0}^{T} v(\mathbf{x}_k(t), \mathbf{u}_k(t)) dt = \int_{t_0}^{T} \langle v(\mathbf{x}(t), .), \omega_t \rangle dt$$

for every function $v \in \mathscr{C}(\mathbf{X} \times \mathbf{U})$. *Then it holds*

$$v^*(t_0, \mathbf{x}_0) = v_R^*(t_0, \mathbf{x}_0)$$

for every $(t_0, \mathbf{x}_0) \in \mathscr{A}$.

Note that this assumption is satisfied under the classical controllability and/or convexity conditions used in the Filippov-Ważewski Theorem with state constraints, see Frankowska and Rampazzo (2000) and the discussions around Assumption I in Gaitsgory and Quincampoix (2009) and Assumption 2 in Henrion and Korda (2014). However, let us point out that Assumption 9.1 does not imply that the infimum is attained in POCP (9.1). Conversely, if the infimum is attained, the values of POCP (9.1) and relaxed POCP (9.2) coincide, and Assumption 9.1 is satisfied.

9.3.2 *Occupation measure*

Given initial data $(t_0, \mathbf{x}_0) \in \mathscr{A}$, and given a relaxed control $\omega_t \in \mathscr{P}(U)$, the unique solution of the ODE

$$\dot{\mathbf{x}}(t) = \langle f(\mathbf{x}(t), .), \omega_t \rangle, \;\; \mathbf{x}(t_0) = \mathbf{x}_0 \tag{9.3}$$

in relaxed POCP (9.2) is given by

$$\mathbf{x}(t) = \mathbf{x}_0 + \int_{t_0}^{t} \langle f(\mathbf{x}(s), .), \omega_s \rangle ds \tag{9.4}$$

for every $t \in [t_0, T]$. Let us then define

$$\mu(dt, d\mathbf{x}, d\mathbf{u}) := dt\, \delta_{\mathbf{x}(t)}(d\mathbf{x})\, \omega_t(d\mathbf{u}) \in \mathscr{M}_+([t_0, T] \times \mathbf{X} \times \mathbf{U}) \tag{9.5}$$

as the occupation measure concentrated uniformly in time on the state trajectory starting at x_0 at time t_0, for the given relaxed control ω_t. An analytic interpretation is that integration w.r.t. the occupation measure is equivalent to time-integration along system trajectories, i.e.

$$\int_{t_0}^{T} v(t, \mathbf{x}(t)) dt = \int_{t_0}^{T} \int_{\mathbf{X}} \int_{\mathbf{U}} v(t, \mathbf{x}) \mu(dt, d\mathbf{x}, d\mathbf{u}) = \langle v, \mu \rangle$$

given any test function $v \in \mathscr{C}([t_0, T] \times \mathbf{X})$.

Introduce the linear operator $\mathcal{L} : \mathscr{C}^1([t_0, T] \times \mathbf{X}) \rightarrow \mathscr{C}([t_0, T] \times \mathbf{X} \times \mathbf{U})$:

$$v \mapsto \mathcal{L}v := \frac{\partial v}{\partial t} + \sum_{i=1}^{n} \frac{\partial v}{\partial x_i} f_i = \frac{\partial v}{\partial t} + \operatorname{grad} v \cdot f.$$

Given a continuously differentiable test function $v \in \mathscr{C}^1([t_0, T] \times \mathbf{X})$, notice that

$$
\begin{aligned}
v(T, \mathbf{x}(T)) - v(t_0, \mathbf{x}(t_0)) &= \int_{t_0}^{T} dv(t, \mathbf{x}(t)) = \int_{t_0}^{T} \dot{v}(t, \mathbf{x}(t))\, dt \\
&= \int_{t_0}^{T} \mathcal{L}v(t, \mathbf{x}(t))\, dt = \langle \mathcal{L}v, \mu \rangle,
\end{aligned}
$$

which can be written more concisely as:

$$
\langle v, \mu_T \rangle - \langle v, \mu_0 \rangle = \langle \mathcal{L}v, \mu \rangle \tag{9.6}
$$

upon defining respectively the initial and terminal occupation measures

$$
\mu_0(dt, d\mathbf{x}) := \delta_{t_0}(dt)\delta_{\mathbf{x}(t_0)}(d\mathbf{x}), \quad \mu_T(dt, d\mathbf{x}) := \delta_T(dt)\delta_{\mathbf{x}(T)}(d\mathbf{x}). \tag{9.7}
$$

Define the adjoint linear operator $\mathcal{L}' : \mathscr{C}([t_0, T] \times \mathbf{X})' \to \mathscr{C}^1([t_0, T] \times \mathbf{X} \times \mathbf{U})'$ by the relation $\langle v, \mathcal{L}'\mu \rangle := \langle \mathcal{L}v, \mu \rangle$ for all $\mu \in \mathscr{M}([t_0, T] \times \mathbf{X})$ and $v \in \mathscr{C}^1([t_0, T] \times \mathbf{X})$. More explicitly, this operator can be expressed as:

$$
\mu \mapsto \mathcal{L}'\mu = -\frac{\partial \mu}{\partial t} - \sum_{i=1}^{n} \frac{\partial(f_i \mu)}{\partial x_i} = -\frac{\partial \mu}{\partial t} - \operatorname{div} f\mu
$$

where the derivatives of measures are understood in the weak sense, i.e. via their action on smooth test functions, and the change of sign comes from integration by parts. Equation (9.6) can be rewritten equivalently as $\langle v, \mu_T \rangle - \langle v, \mu_0 \rangle = \langle v, \mathcal{L}'\mu \rangle$ and since this equation should hold for all test functions $v \in \mathscr{C}^1([t_0, T] \times \mathbf{X})$, we obtain a linear partial differential equation (PDE) on measures $\mathcal{L}'\mu = \mu_T - \mu_0$ that we write:

$$
\frac{\partial \mu}{\partial t} + \operatorname{div} f\mu + \mu_T = \mu_0. \tag{9.8}
$$

This linear transport equation is classical in fluid mechanics, statistical physics and analysis of PDEs. It is called the *equation of conservation of mass*, or the *continuity equation*, or the *advection equation*, or *Liouville's equation*. Under the assumption that the initial data $(t_0, x_0) \in \mathscr{A}$ and the control law $\omega_t \in \mathscr{P}(U)$ are given, the following result can be found e.g. in (Villani, 2003, Theorem 5.34) or Ambrosio (2008).

Lemma 9.1 (Liouville PDE = Cauchy ODE). *There exists a unique solution to the Liouville PDE (9.8) which is concentrated on the solution of the Cauchy ODE (9.3), i.e. such that (9.3.2) and (9.7) hold.*

In our context of conic optimization, the relevance of the Liouville PDE (9.8) is its linearity in the occupation measures μ, μ_0 and μ_T, whereas the Cauchy ODE (9.3) is nonlinear in the state trajectory $\mathbf{x}(t)$.

9.3.3 *Primal LP on measures*

The cost in relaxed POCP (9.2) can therefore be written

$$\int_{t_0}^{T} \langle l(\mathbf{x}(t),.), \omega_t \rangle dt + l_T(\mathbf{x}(T)) = \langle l, \mu \rangle + \langle l_T, \mu_T \rangle$$

and we can now define a relaxed optimal control problem as an LP in the cone of non-negative measures:

$$
\begin{aligned}
p^*(t_0, \mathbf{x}_0) := \min_{\mu, \mu_T} & \left\{ \langle l, \mu \rangle + \langle l_T, \mu_T \rangle : \right. \\
\text{s.t.} \quad & \frac{\partial \mu}{\partial t} + \operatorname{div} f\mu + \mu_T = \delta_{t_0} \delta_{\mathbf{x}_0} \\
& \mu \in \mathscr{M}_+([t_0, T] \times \mathbf{X} \times \mathbf{U}) \\
& \left. \mu_T \in \mathscr{M}_+(\{T\} \times \mathbf{X}_T) \right\},
\end{aligned}
\tag{9.9}
$$

where the minimization is w.r.t. the occupation measure μ (which includes the relaxed control ω_t, see (9.3.2)) and the terminal measure μ_T, for a given initial measure $\mu_0 = \delta_{t_0} \delta_{\mathbf{x}_0}$ which is the right-hand side in the Liouville equation constraint.

Note that in LP (9.9) the infimum is always attained since the admissible set is (weak-star) compact and the functional is linear. However, since classical trajectories are a particular case of relaxed trajectories corresponding to the choice (9.3.2), the minimum in LP (9.9) is smaller than the minimum in relaxed POCP (9.2) (this latter one being equal to the infimum in POCP

(9.1), recall Assumption 9.1), i.e.

$$v^*(t_0, \mathbf{x_0}) \geq p^*(t_0, \mathbf{x_0}). \tag{9.10}$$

The following result, due to Vinter (1993), essentially based on convex duality, shows that there is no gap occurring when considering more general occupation measures than those concentrated on solutions of the ODE.

Lemma 9.2. *It holds* $v^*(t_0, \mathbf{x_0}) = p^*(t_0, \mathbf{x_0})$ *for all* $(t_0, \mathbf{x_0}) \in \mathscr{A}$.

9.3.4   *Dual LP on functions*

Primal measure LP (9.9) has a dual LP in the cone of nonnegative continuous functions:

$$d^*(t_0, \mathbf{x_0}) := \sup_v \{ v(t_0, \mathbf{x_0}) :$$
$$\text{s.t. } l + \tfrac{\partial v}{\partial t} + \text{grad } v \cdot f \in \mathscr{C}_+([t_0, T] \times \mathbf{X} \times \mathbf{U}) \tag{9.11}$$
$$l_T - v(T, .) \in \mathscr{C}_+(\mathbf{X}_T) \},$$

where maximization is with respect to a continuously differentiable function $v \in \mathscr{C}^1([t_0, T] \times X)$ which can be interpreted as a Lagrange multiplier of the Liouville equation in (9.9).

In general the supremum in dual LP (9.11) is not attained, and weak duality with the primal LP (9.9) holds

$$p^*(t_0, \mathbf{x_0}) \geq d^*(t_0, \mathbf{x_0})$$

but it can be shown that there is actually no duality gap:

Lemma 9.3 (No duality gap). *It holds* $p^*(t_0, \mathbf{x_0}) = d^*(t_0, \mathbf{x_0})$ *for all* $(t_0, \mathbf{x_0}) \in \mathscr{A}$.

Proof.   The proof follows along the same lines as the proof of (Henrion and Korda, 2014, Theorem 2). First we observe that $(t_0, \mathbf{x_0}) \in \mathscr{A}$ and Assumption 9.1 imply that $p^*(t_0, \mathbf{x_0})$ is finite. Second, we use the condition that the cone

$$\{ (\langle l, \mu \rangle + \langle l_T, \mu_T \rangle,\ \frac{\partial \mu}{\partial t} + \mathrm{div}\ f\mu + \mu_T) : \mu \in \mathscr{M}_+([t_0, T] \times \mathbf{X} \times \mathbf{U}),$$

$$\mu_T \in \mathscr{M}_+(\{T\} \times \mathbf{X}_T)\}$$

is closed in the weak-star topology. This is a classical sufficient condition for the absence of a duality gap in infinite-dimensional LPs, see e.g. (Barvinok, 2002, Chapter IV, Theorem 7.2). □

9.4 Approximation results

Primal LP (9.9) and dual LP (9.11) are infinite-dimensional conic problems. If we want to solve them with a computer, we invariably have to use discretization and approximation schemes. The aim of this section is to derive various approximation results that prove useful when designing numerical methods based on the Moment-SOS hierarchy.

9.4.1 *Lower bound on value function*

First, observe that there always exists an admissible solution for dual LP (9.11). For example, choose $v(t, \mathbf{x}) := a + b(T - t)$ with $a \in \mathbb{R}$ such that $l_T(\mathbf{x}) \geq a$ on $\mathbf{X}_T$ and $b \in \mathbb{R}$ such that $l(\mathbf{x}, \mathbf{u}) \geq b$ on $\mathbf{X} \times \mathbf{U}$. Moreover, by construction, any admissible function for dual LP (9.11) gives a global lower bound on the value function:

Lemma 9.4 (Lower bound on value function). *If $v \in \mathscr{C}^1([t_0, T] \times \mathbf{X})$ is admissible for dual LP (9.11), then $v^* \geq v$ on $[t_0, T] \times \mathbf{X}$.*

Proof. If $(t_1, \mathbf{x}_1) \notin \mathscr{A}$, then $v^*(t_1, \mathbf{x}_1)$ is unbounded and the statement holds because $v(t_1, \mathbf{x}_1)$ must be finite. Let $(t_1, \mathbf{x}_1) \in \mathscr{A}$ be given. If v is admissible for dual LP (9.11), then for any admissible trajectory $(\mathbf{x}, \mathbf{u}) \in \mathscr{L}^\infty([t_0, T]; \mathbf{X} \times \mathbf{U})$, starting at $\mathbf{x}(t_1) = \mathbf{x}_1$, it holds $\int_{t_1}^T (l(\mathbf{x}(t), \mathbf{u}(t)) + \mathcal{L}v(t, \mathbf{x}(t), \mathbf{u}(t)))dt = \int_{t_1}^T l(\mathbf{x}(t), \mathbf{u}(t))dt + v(T, \mathbf{x}(T)) - v(t_1, \mathbf{x}_1) \geq 0$ since $l + \mathcal{L} \geq 0$ on $[t_0, T] \times \mathbf{X} \times \mathbf{U}$. Moreover $\int_{t_1}^T l(\mathbf{x}(t), \mathbf{u}(t))dt + l_T(\mathbf{x}(T)) \geq \int_{t_1}^T l(\mathbf{x}(t), \mathbf{u}(t))dt + v(T, \mathbf{x}(T))$ since $l_T - v(T, .) \geq 0$ on $\mathbf{X}_T$. Combining the two inequalities yields $\int_{t_1}^{T} l(\mathbf{x}(t), \mathbf{u}(t))dt + l_T(\mathbf{x}(T)) \geq v(t_1, \mathbf{x}_1)$ and the expected inequality follows by taking the infimum over admissible trajectories.   □

The relation between the dual LP (9.11) and the original POCP (9.1) is given by the following result:

Lemma 9.5 (Maximizing sequence). *Given* $(t_0, \mathbf{x}_0) \in \mathscr{A}$, *there is a sequence* $(v_k)_{k \in \mathbb{N}}$ *admissible for the dual LP (9.11) such that* $v^*(t_0, \mathbf{x}_0) \geq v_k(t_0, \mathbf{x}_0)$ *and* $\lim_{k \to \infty} v_k(t_0, \mathbf{x}_0) = v^*(t_0, \mathbf{x}_0)$.

Proof.   From Lemma 9.4, it holds $v^*(t_0, \mathbf{x}_0) \geq v_k(t_0, \mathbf{x}_0)$ for every function v_k admissible for dual LP (9.11). By Assumption 9.1 and Lemma 9.3, it holds $d^*(t_0, \mathbf{x}_0) = p^*(t_0, \mathbf{x}_0) = v^*(t_0, \mathbf{x}_0)$ and hence there exists a maximizing sequence $v_k \in \mathscr{C}^1([t_0, T] \times \mathbf{X})$ for LP (9.11).   □

9.4.2   *Uniform approximation of the value function along trajectories*

In this section, we investigate the properties of maximizing sequences obtained in Lemma 9.5, and in particular their convergence to the value function of POCP (9.1). We first demonstrate the lower semicontinuity of the value of LP (9.9). This leads to the lower semicontinuity of the value of POCP (9.1), by considering Assumption 9.1 and Lemma 9.2. Note that lower semicontinuity is readily ensured when the set $\{(f(\mathbf{x}, \mathbf{u}), l(\mathbf{x}, \mathbf{u}) + a) :$ $\mathbf{u} \in U, a \geq 0\}$ is convex in $\mathbb{R}^{n+1}$ for all $\mathbf{x}$, with U compact, see e.g. (Trélat, 2005, Section 6.2). Indeed, in this case, the infimum is attained in POCP (9.1), and Assumption 9.1 is readily satisfied.

Lemma 9.6 (Lower semi-continuity of the value of the measure LP). *The function* $(t_0, \mathbf{x}_0) \to p^*(t_0, \mathbf{x}_0)$ *is lower semicontinuous.*

Proof.   We need to show that given a sequence $(t_k, \mathbf{x}_k)_{k \in \mathbb{N}}$ such that $\lim_{k \to \infty}(t_k, \mathbf{x}_k) = (t, \mathbf{x}) \in \mathbb{R}^{n+1}$, it holds that $\liminf_{k \to \infty} p^*(t_k, \mathbf{x}_k) \geq p^*(t, \mathbf{x})$. Suppose that $(t, \mathbf{x})$ is such that measure LP (9.9) is feasible. If the left-hand side is not finite, the result holds. If the left-hand side is finite, we can consider, up to taking a subsequence, that

$\liminf_{k\to\infty} p^*(t_k, \mathbf{x}_k) = \lim_{k\to\infty} p^*(t_k, \mathbf{x}_k) < \infty$. Since the infimum is attained in measure LP (9.9), we have a sequence of measures $(\mu_k, \mu_{Tk})_{k\in\mathbb{N}}$ such that $p^*(t_k, \mathbf{x}_k) = \langle \mu_k, l \rangle + \langle \mu_{Tk}, l_T \rangle$ and $\frac{\partial \mu_k}{\partial t} + \operatorname{div} f \mu_k + \mu_{Tk} = \delta_{t_k}\delta_{\mathbf{x}_k}$. Convergence of $(t_k, \mathbf{x}_k)$ to $(t, \mathbf{x})$ implies weak-star convergence of $\delta_{t_k}\delta_{\mathbf{x}_k}$ to $\delta_t\delta_{\mathbf{x}}$. Using the same closedness argument as in the proof of Lemma 9.3, we can consider that, up to a subsequence, μ_k and μ_{Tk} converge to some measures μ and μ_T in the weak-star topology and that $\frac{\partial \mu}{\partial t} + \operatorname{div} f \mu + \mu_T = \delta_t\delta_{\mathbf{x}}$. Hence, we have $\liminf_{k\to\infty} p^*(t_k, \mathbf{x}_k) = \langle \mu, l \rangle + \langle \mu_T, l_T \rangle$ and the pair (μ, μ_T) is feasible for problem $p^*(t, \mathbf{x})$. Therefore $\liminf_{k\to\infty} p^*(t_k, \mathbf{x}_k) \geq p^*(t, \mathbf{x})$ which proves the result when LP (9.9) is feasible for $(t, \mathbf{x})$. Using similar arguments, one can show that if $(t, \mathbf{x})$ is such that LP (9.9) is not feasible, there cannot be infinitely many k such that LP (9.9) is feasible for $(t_k, \mathbf{x}_k)$. $\qquad\square$

The following result extends the convergence properties of the maximizing sequence.

Theorem 9.7 (Uniform convergence along relaxed trajectories).
For any sequence $(v_k)_{k\in\mathbb{N}}$ admissible for the dual LP (9.11), for any solution $(\mathbf{x}, \omega_t)$ of relaxed POCP (9.2), and for any $t \in [t_0, T]$, it holds

$$0 \leq v^*(t, \mathbf{x}(t)) - v_k(t, \mathbf{x}(t)) \leq v^*(t_0, \mathbf{x}_0) - v_k(t_0, \mathbf{x}_0) \underset{k\to\infty}{\to} 0.$$

Proof. Let $(\mathbf{x}_j, u_j)_{j\in\mathbb{N}}$ be an approximating sequence for $(\mathbf{x}, \omega_t)$, whose existence is guaranteed by Assumption 9.1. For any $j \in \mathbb{N}$, $k \in \mathbb{N}$, and $t \in [t_0, T]$, we have $l_T(\mathbf{x}_j(T)) - v_k(T, \mathbf{x}_j(T)) + \int_t^T (l(\mathbf{x}_j(s), u_j(s)) + \mathcal{L}(s, \mathbf{x}_j(s), u_j(s)))ds = l_T(\mathbf{x}_j(T)) + \int_t^T l(\mathbf{x}_j(s), u_j(s))ds - v_k(t, \mathbf{x}_j(t))$. Both the first term and the integrand are positive in the left-hand side. Therefore, the right-hand side is a decreasing function of t. Moreover, the trajectory is suboptimal, and v_k is a lower bound on the value function. It holds that $0 \leq v^*(t, \mathbf{x}_j(t)) - v_k(t, \mathbf{x}_j(t)) \leq l_T(\mathbf{x}_j(T)) + \int_t^T l(\mathbf{x}_j(s), u_j(s))ds - v_k(t, \mathbf{x}_j(t)) \leq l_T(\mathbf{x}_j(T)) + \int_{t_0}^T l(\mathbf{x}_j(s), u_j(s))ds - v_k(t_0, \mathbf{x}_0)$. Letting j tend to infinity, using the lower semicontinuity of v^*, we conclude that $0 \leq v^*(t, \mathbf{x}(t)) -$

$v_k(t, \mathbf{x}(t)) \leq \liminf_{j \to \infty} v^*(t, \mathbf{x}_j(t)) - v_k(t, \mathbf{x}_j(t)) \leq \lim_{j \to \infty}(l_T(\mathbf{x}_j(T)) + \int_{t_0}^{T} l(\mathbf{x}_j(s), \mathbf{u}_j(s))ds - v_k(t_0, \mathbf{x}_0)) = v^*(t_0, \mathbf{x}_0) - v_k(t_0, \mathbf{x}_0).$ □

It is important to notice that Theorem 9.7 holds for any trajectory realizing the minimum of POCP (9.2) and therefore, for all of them simultaneously. In addition, these trajectories are identified with limiting trajectories of POCP (9.1) by Assumption 9.1.

9.5   Optimal control over a set of initial conditions

Liouville equation (9.8) is used as a linear equality constraint in POCP (9.9) with a Dirac right-hand side as an initial condition. However, this right-hand side can be replaced by more general probability measures. The linearity of the constraint allows to extend most of the results of the previous section to this setting. It leads to similar convergence guarantees regarding a (possibly uncountable) set of optimal control problems. These guarantees hold for solutions of a single infinite-dimensional LP.

Suppose that we are given a set of initial conditions $\mathbf{X}_0 \subset \mathbf{X}$, such that $(t_0, \mathbf{x}_0) \in \mathscr{A}$ for every $\mathbf{x}_0 \in \mathbf{X}_0$. Given a probability measure $\xi_0 \in \mathscr{P}(\mathbf{X}_0)$, let:

$$\mu_0(dt, d\mathbf{x}) = \delta_{t_0}(dt)\, \xi_0(d\mathbf{x})$$

and consider the following average value

$$\bar{v}^*(\mu_0) := \int_{\mathbf{X}_0} v^*(t, \mathbf{x})\, \mu_0(dt, d\mathbf{x}) = \langle v^*, \mu_0 \rangle \tag{9.12}$$

where v^* is the value of POCP (9.1). Under Assumption 9.1, by linearity this value is equal to the value of POCP (9.9) with μ_0 as the right-hand side of the equality constraint, namely the primal averaged LP:

$$\bar{p}^*(\mu_0) := \min_{\mu, \mu_T} \{ \langle l, \mu \rangle + \langle l_T, \mu_T \rangle :$$
$$\text{s.t. } \frac{\partial \mu}{\partial t} + \operatorname{div} f\mu + \mu_T = \mu_0$$
$$\mu \in \mathscr{M}_+([t_0, T] \times \mathbf{X} \times \mathbf{U})$$
$$\mu_T \in \mathscr{M}_+(\{T\} \times \mathbf{X}_T)\}, \tag{9.13}$$

with dual averaged LP:

$$\bar{d}^*(\mu_0) := \sup_v \{ \langle v, \mu_0 \rangle :$$
$$\text{s.t. } l + \frac{\partial v}{\partial t} + \text{grad } v \cdot f \in \mathscr{C}_+([t_0, T] \times \mathbf{X} \times \mathbf{U}) \qquad (9.14)$$
$$l_T - v(T, .) \in \mathscr{C}_+(\mathbf{X}_T) \}.$$

The absence of duality gap is justified in the same way as in Lemma 9.3. Moreover, Lemma 9.4 also holds, and, as in Lemma 9.5, we have the existence of maximizing lower bounds v_k such that

$$\lim_{k \to \infty} \langle v_k, \mu_0 \rangle = \bar{v}^*(\mu_0) = \bar{p}^*(\mu_0) = \bar{d}^*(\mu_0).$$

Intuitively, primal LP (9.13) models a superposition of optimal control problems. The LP formulation allows to express it as a single program over measures satisfying a transport equation. A relevant question here is the relation between solutions of averaged measure LP (9.13) and optimal trajectories of the original problem POCP (9.1). The intuition is that measure solutions of LP (9.13) represent a superposition of optimal trajectories of the relaxed POCP (9.2). These trajectories are themselves limiting trajectories of the original POCP (9.1). The superposition principle of (Ambrosio, 2008, Theorem 3.2) allows to formalize this intuition and to extend the result of Theorem 9.7 to this setting.

Theorem 9.8 (Uniform convergence on support of optimal measure). *For any solution* (μ, μ_T) *of primal averaged LP (9.13), there is a parametrized measure* $\xi_t \in \mathscr{P}(\mathbf{X})$ *such that* $\int_U \mu(dt, d\mathbf{x}, d\mathbf{u}) = dt\, \xi_t(d\mathbf{x})$, $\mu_0(dt, d\mathbf{x}) = \delta_{t_0}(dt)\xi_{t_0}(d\mathbf{x})$ *and* $\mu_T(dt, d\mathbf{x}) = \delta_T(dt)\xi_T(d\mathbf{x})$. *In addition, if* $(v_k)_{k \in \mathbb{N}}$ *is a maximizing sequence for dual averaged LP (9.14), for any* $t \in [t_0, T]$, *it holds*

$$0 \le \int_{\mathbf{X}} (v^*(t, \mathbf{x}) - v_k(t, \mathbf{x}))\xi_t(d\mathbf{x}) \le \int_{\mathbf{X}} (v^*(t_0, \mathbf{x}_0) - v_k(t_0, \mathbf{x}_0))\xi_0(d\mathbf{x}_0) \xrightarrow[k \to \infty]{} 0.$$

Proof. The decomposition is given by Lemma 3 in Henrion and Korda (2014). It asserts the existence of a measure $\sigma \in \mathscr{M}(\mathscr{C}([t_0, T], \mathbf{X}))$ supported on trajectories admissible for relaxed POCP (9.2) and such that for any measurable function $w : \mathbf{X} \to \mathbb{R}$, it holds $\int_X w(\mathbf{x})\xi_t(d\mathbf{x}) =$

$\int_{\mathscr{C}([t_0,T],\mathbf{x})} w(\mathbf{x}(t))\sigma(d\mathbf{x}(.))$ for all $t \in [t_0, T]$. By Assumption 9.1, all trajectories of the support of σ are pointwise limits of sequences of feasible trajectories of POCP (9.1). Hence σ-almost all of these sequences must be minimizing sequences for POCP (9.1), otherwise, that would contradict optimality of (μ, μ_T). The result follows by discarding the trajectories which are not limits of minimizing sequences. This does not change σ or ξ_t. Theorem 9.7 applies to σ-almost all these trajectories and we have $0 \leq \int_{\mathscr{C}([t_0,T],\mathbf{x})}(v^*(t,\mathbf{x}(t)) - v_k(t,\mathbf{x}(t)))\sigma(d\mathbf{x}(.)) = \int_X (v^*(t,\mathbf{x}) - v_k(t,\mathbf{x}))\xi_t(d\mathbf{x}) \leq \int_X (v^*(t_0,\mathbf{x}_0) - v_k(t_0,\mathbf{x}_0))\xi_0(d\mathbf{x}_0)$. $\qquad\square$

A remarkable practical implication of this result is that maximizing sequences of averaged dual LP (9.14) provide an approximation to the value function of POCP (9.1) that is uniform in time and almost uniform in space along limits of optimal trajectories starting from $\mathbf{X}_0$.

9.6 A numerical scheme via the Moment-SOS hierarchy

As $\mathbf{X}, \mathbf{X}_T, \mathbf{U}$ are compact, invoking Stone-Weierstrass theorem, we may replace any maximizing sequence of continuous $(v_k)_{k\in\mathbb{N}}$ of (9.11) by a maximizing sequence of polynomials $(p_d)_{d\in\mathbb{N}} \subset \mathbb{R}[\mathbf{x},t]$ of increasing degree. Indeed if l, f, v are polynomials then

$$\theta := l + \frac{\partial v}{\partial t} + \operatorname{grad} v \cdot f \in \mathbb{R}[\mathbf{x},\mathbf{u},t],$$

and therefore the constraint

$$l + \frac{\partial v}{\partial t} + \operatorname{grad} v \cdot f \in \mathscr{C}_+([t_0,T] \times \mathbf{X} \times \mathbf{U}) \tag{9.15}$$

reads $\theta \geq 0$ for all $(\mathbf{x},\mathbf{u},t) \in \mathbf{X} \times \mathbf{U} \times [t_0,T]$. Similarly, the constraint $v(T,\cdot) \leq l_T$ on $\mathbf{X}_T$ is also a positivity constraint on $\mathbf{X}_T$.

Next, assume that $\mathbf{X}$ and $\mathbf{U}$ are the compact basic semi-algebraic sets:

$$\mathbf{X} = \{\mathbf{x} \in \mathbb{R}^n : \quad g_j(\mathbf{x}) \geq 0, \quad j = 1,\ldots,m_x\}$$
$$\mathbf{X}_T = \{\mathbf{x} \in \mathbb{R}^n : \quad r_i(\mathbf{x}) \geq 0, \quad i = 1,\ldots,m_T\}$$
$$\mathbf{U} = \{\mathbf{u} \in \mathbb{R}^m : \quad q_\ell(\mathbf{u}) \geq 0, \quad \ell = 1,\ldots,m_u\},$$

for some polynomials $(g_j, r_i) \subset \mathbb{R}[\mathbf{x}]$, and $(q_\ell) \subset \mathbb{R}[\mathbf{u}]$. Let $d_j :=$ $\lceil \deg(g_j)/2 \rceil$, $j = 1, \ldots, m_x$, $d_j' := \lceil \deg(r_i)/2 \rceil$, $i = 1, \ldots, m_T$, and $d_\ell" :=$ $\lceil \deg(q_\ell)/2 \rceil$, $\ell = 1, \ldots, m_u$. With no loss of generality (and possibly after scaling) we may and will assume that $\mathbf{X}, \mathbf{X}_T \subset [-1, 1]^n$ (resp. $\mathbf{U} \subset [-1, 1]^m$) and $g_1 = n - \|\mathbf{x}\|^2$, $q_1 = m - \|\mathbf{u}\|^2$, $r_1 = g_1$. Moreover let $g_0 := 1$ (a constant polynomial) and $q_0(t) := (T - t)(t - t_0)$. This makes the quadratic modules

$$Q := \{\sum_{j=0}^{m_x} \sigma_j^1 \, g_j + \sum_{\ell=0}^{m_u} \sigma_\ell^2 \, q_\ell : \quad \sigma_j^1, \sigma_\ell^2 \in \Sigma[\mathbf{x}, \mathbf{u}, t]\}$$

and $Q^T := \{\sum_{j=0}^{m_T} \sigma_j^3 \, r_i : \sigma_I^3 \in \Sigma[\mathbf{x}]\}$, archimedean; see §1.4. Let $Q_d \subset Q$ and Q_d^T be their truncated version, i.e.,

$$Q_d := \{\sum_{j=0}^{m_x} \sigma_j^1 \, g_j + \sum_{\ell=0}^{m_u} \sigma_\ell^2 \, q_\ell : \quad \sigma_j^1 \in \Sigma[\mathbf{x}, \mathbf{u}, t]_{d-d_j}, \sigma_\ell^2 \in \Sigma[\mathbf{x}, \mathbf{u}, t]_{d-d_\ell"}\}$$

and $Q_d^T := \{\sum_{i=0}^{m_T} \sigma_i^3 \, r_i : \sigma_i^3 \in \Sigma[\mathbf{x}]_{d-d_i'}\}$. Then one replaces (9.14) with the hierarchy of semidefinite programs indexed by $d \in \mathbb{N}$ and defined by:

$$\rho_d = \sup_{p, \sigma_j^1, \sigma_\ell^2} \{p(t_0, \mathbf{x}_0) : l_T - p(T, \cdot) = \sum_{i=0}^{m_T} \sigma_i^3 \, r_i$$

$$\text{s.t. } l + \frac{\partial p}{\partial t} + \text{grad } p \cdot f = \sum_{j=0}^{m_x} \sigma_j^1 \, g_j + \sum_{\ell=0}^{m_u} \sigma_\ell^2 \, q_\ell \qquad (9.16)$$

$$\sigma_j^1 \in \Sigma[\mathbf{x}, \mathbf{u}, t]_{d-d_j}, \; \sigma_\ell^2 \in \Sigma[\mathbf{x}, \mathbf{u}, t]_{d-d_\ell"}$$

$$\sigma_i^3 \in \Sigma[\mathbf{x}]_{d-d_i'}\}.$$

Obviously (9.16) is a strengthening of (9.11) as one uses Putinar's certificate of positivity (1.17) in lieu of the nonnegativity constraint (9.15).

For each $d \in \mathbb{N}$, (9.16) is the dual of the semidefinite program:

$$\inf_{\mathbf{y}, \mathbf{z}} \{L_\mathbf{y}(l) + L_\mathbf{z}(l_T) :$$

$$\text{s.t. } \mathbf{M}_{d-d_j}(g_j \, \mathbf{y}), \, \mathbf{M}_{d-d"_\ell}(q_\ell \, \mathbf{y}), \, \mathbf{M}_{d-d_i'}(r_i \, \mathbf{z}) \succeq 0,$$

$$\forall j = 0, \ldots, m_x; \; \ell = 0, \ldots, m_u; \; i = 0, \ldots, m_T; \qquad (9.17)$$

$$L_\mathbf{y}\left[l + \frac{\partial (\mathbf{x}^\alpha \, t^k)}{\partial t} + \text{grad } (\mathbf{x}^\alpha \, t^k) \cdot f\right] =$$

$$T^k L_\mathbf{z}(\mathbf{x}^\alpha) - 1_{k=0} \, (\mathbf{x}_0)^\alpha,$$

$$\forall \alpha, k : k + |\alpha| \leq 2d - \deg(f)\}.$$

As the dual (9.16) is a strengthening of (9.11), the primal (9.17) is a *relaxation* of (9.9). Indeed $\mathbf{y}$ (resp. $\mathbf{z}$) is a truncated sequence of pseudo-moments (up to order $2d$) associated with the occupation measure μ (resp. the terminal measure μ_T) in (9.9).

Theorem 9.9. *The optimal value ρ_d of the semidefinite program* (9.16) *satisfies $\rho_d \uparrow v^*(t_0, \mathbf{x}_0)$ as $d \to \infty$.*

Proof. By construction $\rho_d \leq d^*(t_0, \mathbf{x}_0) \leq p^*(t_0, \mathbf{x}_0)$ with p^* and d^* as in (9.9) and (9.11) respectively. In addition by Lemma 9.2 $p^*(t_0, \mathbf{x}_0) = v^*(t_0, \mathbf{x}_0)$. Let $(v_k)_{k \in \mathbb{N}}$ be a maximizing sequence of (9.11). Then for every k their exists $q \in \mathbb{R}[\mathbf{x}, t]$ such that $\|v_k - q\|_\infty < 1/k$ and $\|\nabla(v_k - q)\|_\infty < 1/k$. Let $p := q + (1 + M)(t - 2T)/k$ with $M > n\|f\|_\infty$. Then from $v_k(T, \mathbf{x}) \leq l_T$ on $\mathbf{X}_t$ we deduce that $l_T - p(T, \mathbf{x}) \geq v(T, \mathbf{x}) - q(T, \mathbf{x}) + (1 + M)T/k$ on $\mathbf{X}_T$, and so $l_T - p > 0$ for all $\mathbf{x} \in \mathbf{X}_T$. Moreover,

$$l + \frac{\partial p}{\partial t} + \operatorname{grad} p \cdot f = l + (1 + M)/k + \frac{\partial(q - v + v)}{\partial t} + \operatorname{grad}(q - v + v) \cdot f$$

and therefore

$$l + \frac{\partial p}{\partial t} + \operatorname{grad} p \cdot f > 0, \quad \forall(\mathbf{x}, \mathbf{u}, t) \in \mathbf{X} \times \mathbf{U} \times [t_0, T].$$

By Theorem 1.9 there exists d_k such that $l_T - p(T, \cdot) \in Q_{d_k}^T$ and $l + \frac{\partial p}{\partial t} + \operatorname{grad} p \cdot f \in Q_{d_k}$, that is, p is a feasible solution of (9.16) with $d = d_k$, and with value $v_k(t_0, \mathbf{x}_0) - 1/k \leq \rho_{d_k} \leq p^*(t_0, \mathbf{x}_0) = v^*(t_0, \mathbf{x}_0)$. Next, combining with $v_k(t_0, \mathbf{x}_0) \to v^*(t_0, \mathbf{x}_0)$ as $d \to \infty$, yields the desired result. $\square$

Similarly one may also replace a maximizing sequence $(v_k)_{k \in \mathbb{N}}$ for (9.14) by a maximizing sequence of polynomials $(p_d)_{d \in \mathbb{N}}$ such that $\lim_{d \to \infty} \int p_d d\mu_0 = \int v^* d\mu_0$. One uses the analogue of the semidefinite relaxations (9.16) where now the objective function is $\int p(t_0, \mathbf{x}_0) \, d\mu_0$ instead of $p(t_0, \mathbf{x}_0)$.

9.7 Numerical illustrations

In Sections 9.3 and 9.5 we reformulated nonlinear optimal control problems as abstract linear conic optimization problems that involve manipulations of measures and continuous functions in their full generality. The results presented in Section 9.4 are related to properties of minimizing or max-imizing elements, or sequences of elements for these problems. From a practical point of view, it is possible to construct these sequences using the same numerical tools as in static polynomial optimization. On the primal side, this allows to approximate the minimizing elements of measure LP problems with a converging hierarchy of moment SDP problems. On the dual side, we can construct numerically maximizing sequences of polynomial SOS certificates for the continuous function LP problems. The convergence properties that we investigated hold in particular for these solutions of the Moment-SOS hierarchy.

This section illustrates convergence properties of the sequence of ap-proximations of value functions computed using moment-SOS hierarchies. We consider simple, but largely spread, optimal control problems for which the value function (or optimal trajectories) are known.

9.7.1 *Uniform approximation along an optimal trajectory*

Consider the one-dimensional turnpike POCP analyzed in Section 22.2 of Clarke (2013):

$$
\begin{aligned}
v^*(t_0, x_0) := \inf_u \{ \int_{t_0}^{2} (x(t) + u(t))dt : \\
\text{s.t. } \dot{x}(t) = 1 + x(t) - x(t)u(t) \\
x(t_0) = x_0 \\
u(t) \in [0, 3] \}.
\end{aligned} \tag{9.18}
$$

For this problem, the infimum is attained at a unique optimal control which is piecewise constant. The optimal trajectory $t \mapsto x^*(t)$ starting at $(t_0, x_0) = (0, 0)$ is presented in Figure 9.1. The uniform convergence of approximate value functions $t \mapsto v_k(t, x^*(t))$ to the true value function $t \mapsto v^*(t, x^*(t))$ along this optimal trajectory, stated by Theorem 9.7, is illus-trated in Figure 9.2. Moreover, the difference $t \mapsto v^*(t, x^*(t)) - v_k(t, x^*(t))$ is a decreasing function of time as we observed in the proof of Theorem 9.7.

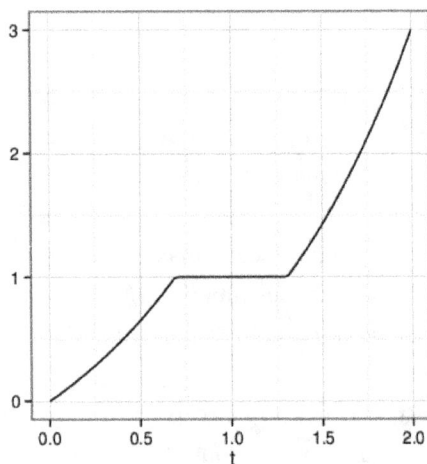

Fig. 9.1 Optimal trajectory $t \mapsto x^*(t)$ starting at $(t_0, x_0) = (0,0)$ for the turnpike POCP (9.18).

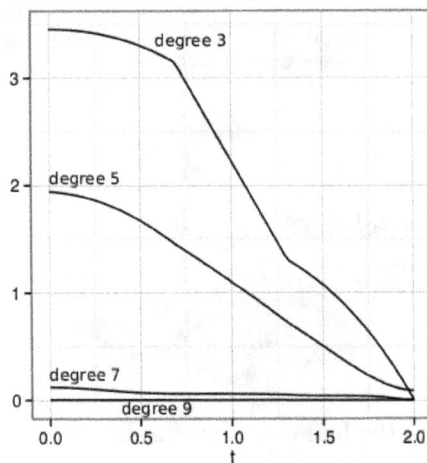

Fig. 9.2 Differences $t \mapsto v^*(t, x^*(t)) - v_k(t, x^*(t))$ between the actual value function and its polynomial approximations of increasing degrees $k = 3, 5, 7, 9$ along the optimal trajectory $t \mapsto x^*(t)$ starting at $(t_0, x_0) = (0,0)$ for the turnpike POCP (9.18). We observe uniform convergence along this trajectory, as well as time decrease of the difference, as predicted by the theory.

9.7.2 *Uniform approximation over a set*

Consider the classical linear quadratic regulator problem:

$$v^*(t_0, x_0) := \inf_u \ \{ \int_0^1 (10\,x(t)^2 + u(t)^2) dt \ :$$
$$\text{s.t. } \dot{x}(t) = x(t) + u(t) \qquad\qquad (9.19)$$
$$x(t_0) = x_0 \ \}.$$

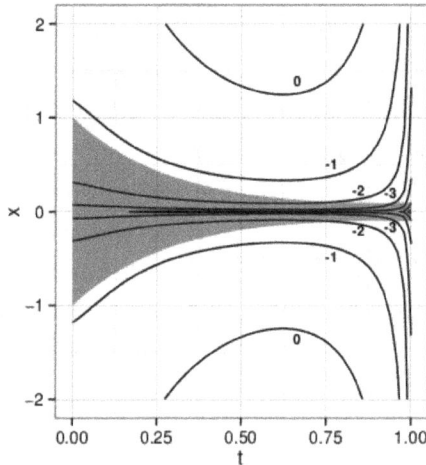

Fig. 9.3 Contour lines (at $0, -1, -2, -3, \ldots$) of the decimal logarithm of the difference $(t, x) \mapsto v^*(t, x) - v_6(t, x)$ between the actual value function and its polynomial approximation of degree 6 for LQR POCP (9.19). The dark area represents the set of optimal trajectories starting from $x_0 \in \mathbf{X}_0 = [-1, 1]$ at time $t_0 = 0$. We observe that the difference is smaller in this area, as predicted by the theory.

For each (t_0, x_0), the infimum is attained and the value of the problem can be computed by solving a Riccati differential equation. To illustrate Theorem 9.8 we are interested in the average value (9.12) for an initial measure μ_0 concentrated at time 0 and uniformly distributed in space in $\mathbf{X}_0 = [-1, 1]$. We approximate this value with primal and dual solutions of LPs (9.13) and (9.14). The contour lines (in decimal logarithmic scale) of the difference between the true value function $(t, x) \mapsto v^*(t, x)$ and a polynomial approximation of degree 6, $(t, x) \mapsto v_6(t, x)$ is represented in Figure 9.3. We also show the support of optimal trajectories starting from

$\mathbf{X}_0$. This illustrates the fact that the approximation of the value function is correct in this region, as stated by Theorem 9.8. It is noticeable that this is computed by a single linear program and provides approximation guarantees uniformly over an uncountable set of optimal control problems.

9.8 Notes and sources

This chapter is mainly from Lasserre et al. (2008). LP formulations of optimal control problems (on ordinary differential equations and partial differential equations) are classical, and can be traced back to the work by L. C. Young, Filippov, as well as Warga and Gamkrelidze, amongst many others. For more details and a historical survey, see e.g. (Fattorini, 1999, Part III). The novelty in Lasserre et al. (2008) is the observation that the infinite-dimensional linear formulations for optimal control problems can be solved numerically with a Moment-SOS hierarchy of the same kind as those used in polynomial optimization Lasserre (2001, 2010, 2015).

Bibliography

Ambrosio, L. (2008). Transport equation and Cauchy problem for non-smooth vector fields, in L. Ambrosio et al. (eds.), *Calculus of Variations and Nonlinear Partial Differential Equations* (Lecture Notes in Mathematics, Vol. 1927, Springer-Verlag, Berlin).

Barvinok, A. (2002). *A Course in Convexity* (American Mathematical Society, Providence, NJ).

Clarke, F. (2013). Functional analysis, Calculus of Variations and Optimal Control (Springer-Verlag, London, UK).

Fattorini, H. O. (1999). *Infinite Dimensional Optimization and Control Theory* (Cambridge Univ. Press, Cambridge, UK).

Gaitsgory, V. and Quincampoix M. (2009). *Linear programming approach to deterministic infinite horizon optimal control problems with discounting*, SIAM J. Control Optim. **48**, pp. 2480–2512.

Frankowska, H. and Rampazzo, F. (2000). *Filippov's and Filippov-Ważewski's theorems on closed domains*, J. Diff. Equations **161**(2), pp. 449–478.

Henrion, D. and Korda, M. (2014). *Convex computation of the region of attraction of polynomial control systems*, IEEE Trans. Aut. Control **59**(2), pp. 297–312.

Lasserre, J. B. (2000). *Optimisation globale et théorie des moments*, C. R. Acad. Sci. Paris, Série I, Math. **331**(11), pp. 929–934.

Lasserre, J. B. (2001). *Global optimization with polynomials and the problem of moments*, SIAM J. Optim. **11**(3), pp. 796–817.

Lasserre, J. B., Henrion, D., Prieur, C., and Trélat, E. (2008). *Nonlinear optimal control via occupation measures and LMI relaxations*, SIAM J. Control Optim. **47**, pp. 1643–1666.

Lasserre, J. B. (2010). *Moments, Positive Polynomials and Their Applications* (Imperial College Press, London, UK).

Lasserre, J. B. (2015). *An Introduction to Polynomial and Semi-Algebraic Optimization* (Cambridge University Press, Cambridge, UK).

Luenberger, D. G. (1969). *Optimization by Vector Space Methods* (Wiley, New York, NY).

Trélat, E. (2005). *Contrôle optimal: Théorie et Applications* (Vuibert, Paris).

Villani, C. (2003). *Topics in Optimal Transportation* (AMS, Providence, RI).

Vinter, R. B. (1993). *Convex duality and nonlinear optimal control*, SIAM J. Control Optim. **31**(2), pp. 518–538.

Chapter 10

Convex Computation of Region of Attraction and Reachable Set

We describe how to apply the Moment-SOS hierarchy to obtain an approximation of the region of attraction and the reachable set for controlled dynamical
systems governed by ordinary differential equations with polynomial dynamics.

10.1 Introduction

Consider the dynamical system:

$$\dot{\mathbf{x}}(t) = f(\mathbf{x}(t), \mathbf{u}(t)), \tag{10.1}$$

where the state $\mathbf{x}(t) \in \mathbb{R}^n$ and control input $\mathbf{u}(t) \in \mathbb{R}^m$ are constrained to
lie in the compact basic semialgebraic sets

$$
\begin{aligned}
\mathbf{x}(t) \in \mathbf{X} &:= \{\mathbf{x} \in \mathbb{R}^n \mid g_i^{\mathbf{X}}(\mathbf{x}) \geq 0, i \in \{1, \dots, n_{\mathbf{X}}\}\}, \, t \in [0, T], \\
\mathbf{u}(t) \in \mathbf{U} &:= \{\mathbf{u} \in \mathbb{R}^m \mid g_i^{\mathbf{U}}(\mathbf{u}) \geq 0, i \in \{1, \dots, n_{\mathbf{U}}\}\}, \, t \in [0, T],
\end{aligned}
\tag{10.2}
$$

with $(g_i^{\mathbf{X}}) \subset \mathbb{R}[\mathbf{x}]$ and $(g_i^{\mathbf{U}}) \subset \mathbb{R}[\mathbf{u}]$. Each component of the vector field f is
assumed to be a polynomial in $(\mathbf{x}, \mathbf{u})$. Given a target set $\mathbf{X}_T$ and an initial
set $\mathbf{X}_I$, the goal of this chapter is to compute the *region of attraction*

$$
\begin{aligned}
\mathbf{X}_0 := \big\{\mathbf{x}_0 \in \mathbf{X} \mid \exists \mathbf{x}(\cdot) \in AC([0, T]; \mathbb{R}^n) \text{ s.t. } \dot{\mathbf{x}}(t) = f(\mathbf{x}(t), \mathbf{u}(t)) \text{ a.e.}, \\
\mathbf{x}(0) = \mathbf{x}_0, \, \mathbf{x}(t) \in \mathbf{X}_T, \, \mathbf{x}(t) \in \mathbf{X} \, \forall t \in [0, T]\big\},
\end{aligned}
$$

(also known as the backward reachable set) and the (forward) reachable
set:

$$\mathbf{X}_R := \{\mathbf{x}_T \in \mathbf{X} \mid \exists \mathbf{x}(\cdot) \in AC([0,T];\mathbb{R}^n) \text{ s.t. } \dot{\mathbf{x}}(t) = f(\mathbf{x}(t),\mathbf{u}(t)) \text{ a.e.,}$$
$$\mathbf{x}(0) \in \mathbf{X}_I, \ \mathbf{x}(t) = \mathbf{x}_T, \ \mathbf{x}(t) \in \mathbf{X} \ \forall t \in [0,T]\}.$$

Here $AC([0,T];\mathbb{R}^n)$ denotes the set of all absolutely continuous functions on $[0,T]$ taking values in $\mathbb{R}^n$ and a.e. stands for "almost everywhere" with respect to the Lebesgue measure on $[0,T]$. The region of attraction (ROA) $\mathbf{X}_0$ is the set of all initial conditions that can be steered to the target $\mathbf{X}_T$ without violating the state constraints using admissible control inputs (i.e., control inputs satisfying $\mathbf{u}(t) \in \mathbf{U}$). The (forward) reachable set $\mathbf{X}_R$ is the set of all states that can be reached at time T starting from the initial set $\mathbf{X}_I$ and using only admissible control inputs.

The target set and the initial set are assumed to be compact basic semialgebraic sets of the form

$$\mathbf{X}_T := \{\mathbf{x} \in \mathbb{R}^n \mid g_i^{\mathbf{X}_T}(\mathbf{x}) \geq 0, \quad i \in \{1,\ldots,n_T\}\} \subset \mathbf{X}$$
$$\mathbf{X}_I := \{\mathbf{x} \in \mathbb{R}^n \mid g_i^{\mathbf{X}_I}(\mathbf{x}) \geq 0, \quad i \in \{1,\ldots,n_I\}\} \subset \mathbf{X},$$

with $(g_i^{\mathbf{X}_T}) \subset \mathbb{R}[\mathbf{x}]$ and $(g_i^{\mathbf{X}_I}) \subset \mathbb{R}[\mathbf{x}]$.

Remark 10.1. Note that the ROA $\mathbf{X}_0$ and the reachable set $\mathbf{X}_R$ are related by a time reversal. Indeed, the reachable set associated to the dynamics $\dot{\mathbf{x}} = f(\mathbf{x},\mathbf{u})$ and initial set $\mathbf{X}_I$ is precisely equal to the ROA associated to the dynamics $\dot{\mathbf{x}} = -f(\mathbf{x},\mathbf{u})$ and target set $\mathbf{X}_T = \mathbf{X}_I$. Therefore, in the rest of this chapter we focus on the ROA problem, with all results holding for the reachable set as well after the exchange of f for $-f$ and $\mathbf{X}_T$ for $\mathbf{X}_I$.

Throughout this chapter we impose the following assumption:

Assumption 10.1.

(a) The polynomials defining the compact basic semialgebraic sets $\mathbf{X}$, $\mathbf{U}$, $\mathbf{X}_T$, $\mathbf{X}_I$ satisfy the Archimedeanity condition. That is, the quadratic modules $Q(\cdot)$ respectively generated by the polynomials $(g_i^{\mathbf{X}})$, $(g_i^{\mathbf{U}})$, $(g_i^{\mathbf{X}_T})$, and $(g_i^{\mathbf{b}\mathbf{X}_I})$ are Archimedean (see Definition 1.3).

(b) For each $\mathbf{x} \in \mathbf{X}$, the set $f(\mathbf{x}, \mathbf{U}) = \{f(\mathbf{x}, \mathbf{u}) \mid \mathbf{u} \in \mathbf{U}\}$ is convex.

10.2   Occupation measures and Liouville equation

In this section we recall the crucial notion of occupation measure already introduced in §9.3.2. This object will be the key ingredient for embedding the *nonlinear* dynamics (10.1) into an infinite-dimensional space of Borel measures where this nonlinear dynamics is equivalently described by a *linear* equation, referred to as the Liouville equation. After that, in Section 10.3, this equation will be used to cast the ROA computation problem as an infinite-dimensional linear program.

The key ingredient in deriving this equation are the initial, occupation and terminal measures generated by ensembles of trajectories of the nonlinear system (10.1). The initial measure μ_0 captures the distribution of the states at time zero; the terminal measure μ_T captures the distribution of the states at time T, after the initial conditions have been transported along the flow of the controlled dynamical system (10.1). The occupation measure captures the evolution of the state and control trajectories in the time interval $(0, T)$. Formally, given an *initial measure* $\mu_0 \in \mathscr{M}(\mathbf{X})_+$ and a set of control trajectories $u(\cdot \mid \mathbf{x}_0)_{\mathbf{x}_0 \in \mathrm{spt}\,\mu_0}$, the *occupation measure* is defined by

$$\mu(A) = \int_{\mathbf{X}} \int_0^T I_A(t, \mathbf{x}(t \mid \mathbf{x}_0), \mathbf{u}(t \mid \mathbf{x}_0)) dt \, d\mu_0(\mathbf{x}_0) \qquad (10.3)$$

whereas the *terminal measure* is defined by

$$\mu_T(B) = \int_{\mathbf{X}} \int_0^T I_B(\mathbf{x}(T \mid \mathbf{x}_0)) \, dt \, d\mu_0(\mathbf{x}_0), \qquad (10.4)$$

for any $A \in \mathscr{B}([0, T] \times \mathbf{X} \times \mathbf{U})$ and $B \in \mathscr{B}(\mathbf{X})$, where $\mathbf{x}(\cdot \mid \mathbf{x}_0)$ denotes the unique trajectory of (10.1) associated to the control input $\mathbf{u}(\cdot \mid \mathbf{x}_0)$. These measures satisfy the following important property: given any functions

$h \in L_\infty([0, T] \times \mathbf{X} \times \mathbf{U})$ and $h_T \in L_\infty(\mathbf{X})$, it holds

$$\int_{\mathbf{X}} \int_0^T h(t, \mathbf{x}(t \mid \mathbf{x}_0), \mathbf{u}(t \mid \mathbf{x}_0))\, dt\, d\mu_0(\mathbf{x}_0) = \int_{[0,T] \times \mathbf{X} \times \mathbf{U}} h(t, \mathbf{x}, \mathbf{u})\, d\mu(t, \mathbf{x}, \mathbf{u})$$

(10.5)

and

$$\int_{\mathbf{X}} h_T(\mathbf{x}(T \mid \mathbf{x}_0))\, d\mu_0(\mathbf{x}_0) = \int_{\mathbf{X}} h_T(\mathbf{x})\, d\mu_T(\mathbf{x}).$$ (10.6)

In other words, for the occupation measure μ, the temporal integration along the trajectories of (10.1) is replaced by spatial integration with respect to μ. This fact will be crucial for obtaining a linear representation of the nonlinear dynamics. In order to derive this representation, consider any test function $v \in \mathscr{C}^1([0, T] \times \mathbf{X})$; then for any initial condition $\mathbf{x}_0 \in \mathrm{spt}\,\mu_0$, we get

$$\begin{aligned} v(T, \mathbf{x}(T \mid \mathbf{x}_0)) - v(0, \mathbf{x}_0) &= \int_0^T \frac{d}{dt} v(t, \mathbf{x}(t \mid \mathbf{x}_0))\, dt \\ &= \int_0^T [\, \frac{\partial v}{\partial t} + \frac{\partial v}{\partial \mathbf{x}} \cdot f(\mathbf{x}(t \mid \mathbf{x}_0), \mathbf{u}(t \mid \mathbf{x}_0))\,]\, dt, \end{aligned}$$

where $\cdot$ denotes the standard dot product of vectors. Integrating with respect to μ_0 and using (10.5) and (10.6), we get

$$\int_{\mathbf{X}} v(T, \mathbf{x})\, d\mu_T(\mathbf{x}) = \int_{\mathbf{X}} v(0, \mathbf{x})\, d\mu_0(\mathbf{x})$$ (10.7)

$$+ \int_{[0,T] \times \mathbf{X} \times \mathbf{U}} [\, \frac{\partial v}{\partial t} + \frac{\partial v}{\partial \mathbf{x}} \cdot f\,]\, d\mu(t, \mathbf{x}, \mathbf{u}).$$

Equation (10.7) is referred to as *Liouville equation*; see also (9.6). Notice that this equation is *linear* in the triplet of measures (μ_0, μ, μ_T), irrespective of whether the vector field f is linear or not. This equation therefore represents a linear *embedding* of the non-linear dynamical system.

10.3 Infinite dimensional LP characterization of ROA

In this section we give a characterization of the ROA as a solution to an infinite-dimensional linear program (LP). Let the linear operator $\mathcal{L}$: $\mathscr{C}^1([0,T] \times \mathbf{X}) \to \mathscr{C}([0,T] \times \mathbf{X} \times \mathbf{U})$ be given by

$$v \mapsto \mathcal{L}v := \frac{\partial v}{\partial t} + \frac{\partial v}{\partial \mathbf{x}} \cdot f.$$

The infinite-dimensional LP then reads:

$$
\begin{aligned}
p^\star = \sup_{\mu,\mu_0,\mu_T} \{ &\int_{\mathbf{X}} 1 \, d\mu_0 \quad \text{s.t.} \quad \mu_0 \leq \lambda, \\
&\int_{\mathbf{X}_T} v(T,\cdot) \, d\mu_T = \int_{\mathbf{X}} v(0,\cdot) \, d\mu_0 + \int_{[0,T] \times \mathbf{X} \times \mathbf{U}} \mathcal{L}v \, d\mu, \\
&\forall v \in \mathscr{C}^1([0,T] \times \mathbf{X}); \\
&\mu \in \mathscr{M}([0,T] \times \mathbf{X} \times \mathbf{U})_+, \\
&\mu_0 \in \mathscr{M}(\mathbf{X})_+, \ \mu_T \in \mathscr{M}(\mathbf{X}_T)_+ \},
\end{aligned}
\tag{10.8}
$$

where λ denotes the Lebesgue measure.

Interpretation. The equality constraint of (10.8) is nothing but the Liouville equation (10.7), which is a relaxed representation of the nonlinear dynamics (10.1) in terms of the initial, final and occupation measures μ_0, μ_T and μ. The conic inclusions encode nonnegativity of the measures as well as appropriate support constraints which in turn enforce satisfaction of the constraints (10.2) and of the requirement $\mathbf{x}(t) \in \mathbf{X}_T$. The constraint $\mu_0 \leq \lambda$ is equivalent to the requirement that the initial measure μ_0 is absolutely continuous with respect to the Lebesgue measure λ with density less than or equal to one almost everywhere. The objective function then simply maximizes the mass of μ_0; intuitively speaking, we are trying to make the density of the initial measure μ_0 as large as possible while bounded by one and while satisfying the Liouville equation as well as all support constraints. It turns out that the μ_0 component of any optimal solution to (10.8) has the very simple form of

$$d\mu_0(\mathbf{x}) = I_{\mathbf{X}_0}(\mathbf{x})d\mathbf{x},$$

where $I_{\mathbf{X}_0}(\mathbf{x})$ is the indicator function of the ROA $\mathbf{X}_0$. Therefore also $p^\star = \lambda(\mathbf{X}_0) = \text{vol}\,\mathbf{X}_0$. This is summarized by the following theorem.

> **Theorem 10.2.** *If Assumption 10.1 holds, then $p^\star = \text{vol}\,\mathbf{X}_0$. Moreover, the supremum in (10.8) is attained and by a measure μ_0 which is the restriction of the Lebesgue measure to the ROA $\mathbf{X}_0$.*

10.3.1 *Dual LP*

In this section we formulate the LP dual to (10.8). The LP is derived using standard infinite-dimensional LP duality theory (e.g., Anderson & Nash (1987)) after converting the inequality constraint $\mu_0 \leq \lambda$ to an equality constraint $\mu_0 + \hat{\mu}_0 = \lambda$ with $\hat{\mu}_0 \in \mathscr{M}(\mathbf{X})_+$. The dual LP then reads

$$
\begin{aligned}
d^\star = \inf_{v \in \mathscr{C}^1([0,T] \times \mathbf{X}),\, w \in \mathscr{C}(\mathbf{X})} & \int_{\mathbf{X}} w(\mathbf{x})\, d\lambda(\mathbf{x}) \\
\text{s.t.} \qquad\qquad -\mathcal{L}v &\in \mathscr{C}([0,T] \times \mathbf{X} \times \mathbf{U})_+, \\
w - v(0,\cdot) - 1 &\in \mathscr{C}(\mathbf{X})_+, \\
v(T,\cdot) &\in \mathscr{C}(\mathbf{X}_T)_+, \\
w &\in \mathscr{C}(\mathbf{X})_+.
\end{aligned}
$$

$$(10.9)$$

Interpretation. The condition $-\mathcal{L}v \in \mathscr{C}_+([0,T] \times \mathbf{X} \times \mathbf{U})$ implies that the function v should decrease along any trajectory of (10.1) satisfying the constraints (10.2). Combining this with the constraint that $v(T,\cdot) \in \mathscr{C}(\mathbf{X}_T)_+$ implies that $v(0,\mathbf{x}) \geq 0$ for all $x \in \mathbf{X}_0$. As a result, due to the third and fourth constraints of (10.9), we have $w(\mathbf{x}) \geq I_{\mathbf{X}_0}(\mathbf{x})$ for all $x \in \mathbf{X}$.

Therefore necessarily

$$\int_{\mathbf{X}} w(\mathbf{x})\, d\lambda(\mathbf{x}) \geq \operatorname{vol} \mathbf{X}_0 \quad \text{and} \quad \{\mathbf{x} \mid v(0,\mathbf{x}) \geq 0\} \supset \mathbf{X}_0$$

for any w feasible in (10.9). In other words, any v feasible in (10.9) provides us with an *outer approximation* of the ROA $\mathbf{X}_0$ and the objective functional of (10.9), along with the inequalities $w \geq I_{\mathbf{X}_0}$ and $w \geq 1 + v(0,\cdot)$, can be viewed as a convex proxy for the minimization of $\operatorname{vol}\{\mathbf{x} \mid v(0,\mathbf{x}) \geq 0\}$ (which is a non-convex function of v). This is summarized in the following lemma:

Lemma 10.3. *Let (v,w) be a pair of function feasible in (10.9). Then $v(0,\cdot) \geq 0$ on $\mathbf{X}_0$, $w \geq 1$ on $\mathbf{X}_0$ and $w \geq I_{\mathbf{X}_0}$ on $\mathbf{X}$. In particular $\{\mathbf{x} \mid v(0,\mathbf{x}) \geq 0\} \supset \mathbf{X}_0$.*

Proof. By definition of $\mathbf{X}_0$, given any $\mathbf{x}_0 \in \mathbf{X}_0$ there exists $\mathbf{u}(t)$ such that $\mathbf{x}(t) \in \mathbf{X}$, $\mathbf{u}(t) \in \mathbf{U}$ for all $t \in [0,T]$ and $\mathbf{x}(t) \in \mathbf{X}_T$. Therefore, since $v(T,\cdot) \geq 0$ on $\mathbf{X}_T$ and $\mathcal{L}v \leq 0$ on $[0,T] \times \mathbf{X} \times \mathbf{U}$, we get

$$0 \leq v(T, \mathbf{x}(t)) = v(0, \mathbf{x}_0) + \int_0^T \frac{d}{dt} v(t, \mathbf{x}(t))\, dt$$

$$= v(0, \mathbf{x}_0) + \int_0^T \mathcal{L}v(t, \mathbf{x}(t), \mathbf{u}(t))\, dt$$

$$\leq v(0, \mathbf{x}_0) \leq w(\mathbf{x}_0) - 1,$$

where the last inequality follows from the second constraint of (10.9). The fact that $w \geq I_{\mathbf{X}_0}$ follows from the last constraint of (10.9) which requires w to be nonnegative on $\mathbf{X}$. $\qquad\square$

The following results is crucial for subsequent develops; it states that there is no duality gap between the LPs (10.8) and (10.9).

Theorem 10.4. *There is no duality gap between the primal infinite-dimensional LP problem (10.8) and the infinite-dimensional dual LP problem (10.9), i.e., $p^\star = d^\star$.*

Proof.   Let

$$\mathcal{M} := \mathscr{M}([0,T] \times \mathbf{X} \times \mathbf{U})_+ \times \mathscr{M}(\mathbf{X})_+ \times \mathscr{M}(\mathbf{X}_T)_+ \times \mathscr{M}(\mathbf{X})_+,$$

$$\mathcal{C} := \mathscr{C}([0,T] \times \mathbf{X} \times \mathbf{U})_+ \times \mathscr{C}(\mathbf{X})_+ \times \mathscr{C}(\mathbf{X}_T)_+ \times \mathscr{C}(\mathbf{X})_+,$$

and let $\mathcal{K}$ and $\mathcal{K}^\star$ denote the positive cones of $\mathcal{M}$ and $\mathcal{C}$ respectively. Note that the cone $\mathcal{K}$ of nonnegative measures of $\mathcal{M}$ can be identified with the topological dual of the cone $\mathcal{K}$ of nonnegative continuous functions of $\mathcal{C}$. The cone $\mathcal{K}$ is equipped with the weak-$\star$ topology. After replacing the constraint $\mu_0 \leq \lambda$ with $\mu_0 + \hat{\mu}_0 = \lambda$, $\hat{\mu}_0 \in \mathscr{M}(\mathbf{X})_+$, the LP problem (10.8) can be rewritten as

$$p^\star = \sup_{\gamma} \{ \langle \gamma, c \rangle : \mathcal{A}\gamma = \beta; \gamma \in \mathcal{K} \}, \tag{10.10}$$

where the infimum is over the vector $\gamma := (\mu, \mu_0, \mu_T, \hat{\mu}_0)$, the linear operator $\mathcal{A} : \mathcal{K} \to \mathscr{C}^1([0,T] \times \mathbf{X})^\star \times M(\mathbf{X})$ is defined by

$$\mathcal{A}\gamma := (-\mathcal{L}^\star\mu - \delta_0 \otimes \mu_0 + \delta_T \otimes \mu_T \ , \ \mu_0 + \hat{\mu}_0),$$

where $\mathcal{L}$ is the adjoint of the operator $\mathcal{L}$, the right-hand side of the equality constraint in (10.10) is the vector of measures $\beta := (0, \lambda) \in \mathscr{M}([0,T] \times \mathbf{X}) \times \mathscr{M}(\mathbf{X})$, the vector function in the objective is $c := (0,1,0,0) \in \mathcal{C}$, so the objective function itself is

$$\langle \gamma, c \rangle = \int_{\mathbf{X}} 1 \, d\mu_0 = \mu_0(\mathbf{X}).$$

The dual LP to (10.10) reads

$$d^\star = \inf_z \{ \langle \beta, z \rangle : \mathcal{A}^\star(z) - c \in \mathcal{K}^\star \}, \tag{10.11}$$

where the infimum is over $z := (v, w) \in \mathscr{C}^1([0,T] \times \mathbf{X}) \times \mathscr{C}(\mathbf{X})$, and the linear operator $\mathcal{A}^\star : \mathscr{C}^1([0,T] \times \mathbf{X}) \times \mathscr{C}(\mathbf{X}) \to \mathcal{C}$ is defined by

$$\mathcal{A}^\star z := (-\mathcal{L}v, \ w - v(0, \cdot), \ v(T, \cdot), \ w)$$

and satisfies the adjoint relation $\langle \mathcal{A}\gamma, z \rangle = \langle \gamma, \mathcal{A}^\star z \rangle$. The LP problem (10.11) is exactly the LP problem (10.9).

To conclude the proof we use an argument similar to that of (Lasserre, 2009, Section C.4). From (Anderson & Nash, 1987, Theorem 3.10) there is no duality gap between LPs (10.10) and (10.11) if the supremum $p^\star$ is finite and the set $S := \{(\mathcal{A}\gamma, \langle \gamma, c \rangle) : \gamma \in \mathcal{K}\}$ is closed in the weak-* topology of $\mathscr{C}^1([0,T] \times \mathbf{X})^\star \times \mathscr{M}(\mathbf{X}) \times \mathbb{R}$. The fact that $p^\star$ is finite follows readily from the constraint $\mu_0 + \hat{\mu}_0 = \lambda$, $\hat{\mu}_0 \geq 0$, and from compactness of $\mathbf{X}$. To prove closedness, we first remark that $\mathcal{A}$ is weakly-* continuous[1] since $\mathcal{A}^\star(z) \in \mathcal{C}$ for all $z \in \mathscr{C}^1([0,T] \times \mathbf{X}) \times \mathscr{C}(\mathbf{X})$. Then we consider a sequence $\gamma_k = (\mu^k, \mu_0^k, \mu_T^k, \hat{\mu}_0^k) \in \mathcal{K}$ and we want to show that $(\nu, a) := \lim_{k \to \infty}(\mathcal{A}\gamma_k, \langle \gamma_k, c \rangle)$ belongs to S, where $\nu \in \mathscr{C}^1([0,T] \times \mathbf{X})^\star \times \mathscr{M}(\mathbf{X})$ and $a \in \mathbb{R}$. To this end, consider first the test function $z_1 = (T-t, 1)$ which gives $\langle \mathcal{A}\gamma_k, z_1 \rangle = \mu^k(0, T \times \mathbf{X} \times \mathbf{U}) + \mu_0^k(\mathbf{X}) + \hat{\mu}_0^k(\mathbf{X}) \to \langle \nu, z_1 \rangle < \infty$; since the measures are nonnegative, this implies that the sequences of measures μ^k, μ_0^k and $\hat{\mu}_0^k$ are bounded. Next, taking the test function $z_2 = (1, 1)$ gives $\langle \mathcal{A}\gamma_k, z_2 \rangle = \mu_T(\mathbf{X}) + \hat{\mu}_0(\mathbf{X}) \to \langle \nu, z_2 \rangle < \infty$; this implies that the sequence μ_T^k is bounded as well. Thus, from the weak-* compactness of the unit ball (Banach-Alaoglu theorem), there is a subsequence γ_{k_i} that converges weakly-* to an element $\gamma \in \mathcal{K}$ so that $\lim_{i \to \infty}(\mathcal{A}\gamma_{k_i}, \langle \gamma_{k_i}, c \rangle) \in S$ by continuity of $\mathcal{A}$. □

This theorem allows us to prove the following result, which will be important for establishing convergence of outer approximations in Section 10.4.

Theorem 10.5. *There is a sequence of feasible solutions to the dual LP (10.9) such that its w-component converges from above to $I_{\mathbf{X}_0}$ in L^1 norm.*

Proof. By Theorem 10.2, the optimal solution to the primal is attained by the restriction of the Lebesgue measure to $\mathbf{X}_0$. Consequently,

$$p^\star = \int_{\mathbf{X}} I_{\mathbf{X}_0}(\mathbf{x}) d\lambda(\mathbf{x}). \tag{10.12}$$

By Theorem 10.4, there is no duality gap ($p^\star = d^\star$), and therefore there exists a sequence $(v_k, w_k) \in \mathscr{C}^1([0,T] \times \mathbf{X}) \times \mathscr{C}(\mathbf{X})$ feasible in (10.9) such

---

[1] The weak-* topology on $\mathscr{C}^1([0,T] \times \mathbf{X})^\star \times \mathscr{M}(\mathbf{X})$ is induced by the standard topologies on $\mathscr{C}^1$ and $\mathscr{C}$ — the topology of uniform convergence of the function and its derivative on $\mathscr{C}^1$ and the topology of uniform convergence on $\mathscr{C}$.

that

$$p^\star = d^\star = \lim_{k \to \infty} \int_{\mathbf{X}} w_k(\mathbf{x}) \, d\lambda(\mathbf{x}). \qquad (10.13)$$

From Lemma 10.3 we have $w_k \geq I_{\mathbf{X}_0}$ on $\mathbf{X}$ for all k. Thus, subtracting (10.12) from (10.13) gives

$$\lim_{k \to \infty} \int_{\mathbf{X}} (w_k(\mathbf{x}) - I_{\mathbf{X}_0}(x)) \, d\lambda(\mathbf{x}) = 0,$$

where the integrand is nonnegative. Hence w_k converges to $I_{\mathbf{X}_0}$ in L^1 norm. $\qquad\qquad\square$

10.4 Finite-dimensional SDP approximations

In this section we derive finite-dimensional semidefinite programming (SDP) approximations of the infinite-dimensional LPs (10.8) and (10.9).

The inclusions in the cones $\mathscr{M}(\cdot)_+$ and $\mathscr{C}(\cdot)_+$ are approximated by inclusions in the truncated moment cone (respectively the truncated quadratic module) as described in Chapter 1. In addition to this, we truncate the equality constraints

$$\int_{\mathbf{X}_T} v(T, \mathbf{x}) \, d\mu_T(\mathbf{x}) - \int_{\mathbf{X}} v(0, \mathbf{x}) \, d\mu_0(\mathbf{x}) - \int_{[0,T] \times \mathbf{X} \times \mathbf{U}} \mathcal{L}v(t, \mathbf{x}, \mathbf{u}) \, d\mu(t, \mathbf{x}, \mathbf{u}) = 0,$$

$$\int_{\mathbf{X}} w(\mathbf{x}) \, d\mu_0(\mathbf{x}) + \int_{\mathbf{X}} w(\mathbf{x}) \, d\hat{\mu}_0(\mathbf{x}) = \int_{\mathbf{X}} w(\mathbf{x}) \, d\lambda(\mathbf{x})$$

by enforcing these only for the particular choice of test functions $v(t, \mathbf{x}) = t^\alpha \mathbf{x}^\beta$ and $w(\mathbf{x}) = \mathbf{x}^\gamma$ for all $\alpha \in \mathbb{N}$, $\beta \in \mathbb{N}^n$ and $\gamma \in \mathbb{N}^n$ such that $\alpha + |\beta| \leq d_v$ and $\gamma \leq 2d$, where

$$d_v := 2d - \deg f + 1.$$

The resulting finite-dimensional truncation of this linear system of equations is denoted by

$$\mathbf{A}_d(\mathbf{y}, \mathbf{y}_0, \mathbf{y}_T, \hat{\mathbf{y}}_0) = b_d,$$

where $\mathbf{y}$, $\mathbf{y}_0$, $\mathbf{y}_T$ and $\hat{\mathbf{y}}_0$ represent the truncated moment sequences of the measures $(\mu, \mu_0, \mu_T, \hat{\mu}_0)$.

Recall that $g_0(\cdot) = 1$. The finite-dimensional SDP relaxation of the primal LP (10.8) reads:

$$p_d^\star = \max_{(\mathbf{y}, \mathbf{y}_0, \mathbf{y}_T, \hat{\mathbf{y}}_0)} \{ (\mathbf{y}_0)_0 : \mathbf{A}_d(\mathbf{y}, \mathbf{y}_0, \mathbf{y}_T, \hat{\mathbf{y}}_0) = b_d;$$
$$\mathbf{M}_d((T-t)t\,\mathbf{y}), \mathbf{M}_d(g_i^{\mathbf{X}}\,\mathbf{y})\,\mathbf{M}_d(g_j^{\mathbf{U}}\,\mathbf{y}) \succeq 0,$$
$$\forall\, 0 \leq i \leq n_{\mathbf{X}},\ 0 \leq j \leq n_{\mathbf{U}}; \tag{10.14}$$
$$\mathbf{M}_d(g_i^{\mathbf{X}}\,\mathbf{y}_0), \mathbf{M}_d(g_i^{\mathbf{X}}\,\hat{\mathbf{y}}_0), \mathbf{M}_d(g_j^{\mathbf{X}_T}\,\hat{\mathbf{y}}_T) \succeq 0,$$
$$\forall\, 0 \leq i \leq n_{\mathbf{X}},\ 0 \leq j \leq n_{\mathbf{X}_T} \}.$$

Denote by $Q_d((\cdot))$ the truncated quadratic module generated by the family of polynomials $(\cdots)$; see (1.5). The dual tightening, which will provide us with a sequence of converging outer approximations, reads:

$$d_d^\star = \inf_{v,w} \{ \mathbf{w}^\top \mathbf{1} : -\mathcal{L}v \in Q_d((T-t)t, (g_i^{\mathbf{X}}), (g_j^{\mathbf{U}}));$$
$$w - v(0, \cdot) - 1 \in Q_d((g_i^{\mathbf{X}})),$$
$$v(T, \cdot) \in Q_d((g_i^{\mathbf{X}_T})),\ w \in Q_d((g_i^{\mathbf{X}})), \tag{10.15}$$
$$v \in \mathbb{R}[t, \mathbf{x}]_{d_v},\ w \in \mathbb{R}[\mathbf{x}]_{2d} \},$$

where $\mathbf{w}$ denotes the coefficient vector of the polynomial w and $\mathbf{1}$ the vector of moments of the Lebesgue measure on $\mathbf{X}$, both indexed in the same basis.

10.4.1 *Outer approximation*

The solutions to the SDP (10.15) indexed by the degree d provide us with a sequence of outer approximations to the ROA $\mathbf{X}_0$. Let $(w_d, v_d) \in \mathbb{R}[\mathbf{x}] \times \mathbb{R}[t, \mathbf{x}]$ denote a solution to (10.15). Then we define

$$\mathbf{X}_{0d} := \{\mathbf{x} \in \mathbf{X} : v_d(0, \mathbf{x}) \geq 0\}. \tag{10.16}$$

Since any v_d feasible in (10.15) is feasible in the infinite-dimensional dual LP (10.9), Lemma 10.3 implies that

$$\mathbf{X}_{0d} \supset \mathbf{X}_0,$$

i.e. $\mathbf{X}_{0d}$ is an outer approximation of $\mathbf{X}_0$. Next, we establish convergence of these outer approximations to $\mathbf{X}_0$, in the sense of volume discrepancy tending to zero. In order to do so, we need the following lemma:

Lemma 10.6. *Let $w_d \in \mathbb{R}_{2d}[\mathbf{x}]$ denote the w-component of an optimal solution to the dual SDP (10.15). Then w_d converges from above to $I_{\mathbf{X}_0}$ in L^1 norm, i.e.,*

$$\lim_{d\to\infty} \int_{\mathbf{X}} w_d(\mathbf{x}) - I_{\mathbf{X}_0}\, d\mathbf{x} = 0.$$

Proof. From Lemma 10.3 and Lemma 10.5, for every $\epsilon > 0$ there exists a $(v, w) \in C^1([0, T] \times \mathbf{X}) \times C(\mathbf{X})$ feasible in (10.9) such that $w \geq I_{\mathbf{X}_0}$ and $\int_{\mathbf{X}}(w - I_{\mathbf{X}_0})\, d\lambda < \epsilon$. Set

$$\tilde{v}(t, \mathbf{x}) := v(t, \mathbf{x}) - \epsilon t + (T + 1)\epsilon,$$
$$\tilde{w}(\mathbf{x}) := w(\mathbf{x}) + (T + 3)\epsilon.$$

Since v is feasible in (10.9), we have $\mathcal{L}\tilde{v} = \mathcal{L}v - \epsilon$, and $\tilde{v}(T, \mathbf{x}) = v(T, \mathbf{x}) + \epsilon$. Since also $\tilde{w}(\mathbf{x}) - \tilde{v}(0, \mathbf{x}) \geq 1 + 2\epsilon$, it follows that $(\tilde{v}, \tilde{w})$ is *strictly* feasible in (10.9) with a margin at least ϵ. Since $[0, T] \times \mathbf{X}$ and $\mathbf{X}$ are compact, there exist[2] polynomials $\hat{v}$ and $\hat{w}$ of a sufficiently high degree such that $\sup_{[0,T]\times\mathbf{X}} |\tilde{v} - \hat{v}| < \epsilon$, $\sup_{[0,T]\times\mathbf{X}\times\mathbf{U}} |\mathcal{L}\tilde{v} - \mathcal{L}\hat{v}| < \epsilon$ and $\sup_{\mathbf{X}} |\tilde{w} - \hat{w}| < \epsilon$. The pair of polynomials $(\hat{v}, \hat{w})$ is therefore *strictly* feasible in (10.9) and as a result, under Assumption 10.1, feasible in (10.15) for a sufficiently large degree d (this follows from the Putinar positivestellensatz 1.9), and moreover $\hat{w} \geq w$. Consequently, $\int_{\mathbf{X}} |\tilde{w} - \hat{w}|\, d\lambda \leq \epsilon\lambda(\mathbf{X})$, and so $\int_{\mathbf{X}}(\hat{w} - w)\, d\lambda \leq \epsilon\lambda(\mathbf{X})(T + 4)$. Therefore

$$\int_{\mathbf{X}}(\hat{w} - I_{\mathbf{X}_0})\, d\lambda < \epsilon K, \quad \hat{w} \geq I_{\mathbf{X}_0},$$

where $K := [1 + (T + 4)\lambda(\mathbf{X})] < \infty$ is a constant. This proves the first statement since ϵ was arbitrary. $\square$

The following establishes convergence of these outer approximations as the degree d tends to infinity.

---

[2]This follows from an extension of the Stone-Weierstrass theorem that allows for a simultaneous uniform approximation of a function and its derivatives by a polynomial on a compact set; see, e.g., Bagby et al. (2002).

Theorem 10.7. *Let* $(v_d, w_d) \in \mathbb{R}_{d_v}[t, \mathbf{x}] \times \mathbb{R}_{2d}[\mathbf{x}]$ *denote a solution to the dual SDP problem (10.15). The sets* $\mathbf{X}_{0d}$ *defined in (10.16) converge to the ROA* $\mathbf{X}_0$ *from the outside, i.e.,*

$$\mathbf{X}_{0d} \supset \mathbf{X}_0 \quad \forall d \in \mathbb{N} \quad and \quad \lim_{d \to \infty} \text{vol}(\mathbf{X}_{0d} \setminus \mathbf{X}_0) = 0.$$

Proof. The inclusion $\mathbf{X}_{0d} \supset \mathbf{X}_0$ follows from Lemma 10.3 since any solution to (10.15) is feasible in (10.9). Next, from Lemma 10.3 we have $w_d \geq I_{\mathbf{X}_0}$ and therefore, since $w_d \geq v_d(0, \cdot) + 1$ on $\mathbf{X}$, we have $w_d \geq I_{\mathbf{X}_{0d}} \geq I_{\mathbf{X}_0}$ on $\mathbf{X}$. In addition, from Lemma 10.6 we have $w_d \to I_{\mathbf{X}_0}$ in L^1 norm on $\mathbf{X}$; therefore

$$\text{vol}(\mathbf{X}_0) = \lambda(\mathbf{X}_0) = \int_{\mathbf{X}} I_{\mathbf{X}_0} \, d\lambda = \lim_{d \to \infty} \int_{\mathbf{X}} w_d \, d\lambda \geq \lim_{d \to \infty} \int_{\mathbf{X}} I_{\mathbf{X}_{0d}} \, d\lambda$$
$$= \lim_{d \to \infty} \lambda(\mathbf{X}_{0d}).$$

But since $\mathbf{X}_0 \subset \mathbf{X}_{0d}$ we must have $\lambda(\mathbf{X}_0) \leq \lambda(\mathbf{X}_{0d})$ and the theorem follows. $\qquad\qquad\qquad\square$

10.5 Numerical examples

10.5.1 *Van der Pol oscillator*

Consider a scaled version of the uncontrolled reversed-time Van der Pol oscillator given by:

$$\dot{x}_1 = -2\,x_2; \quad \dot{x}_2 = 0.8\,x_1 + 10\,(x_1^2 - 0.21)\,x_2.$$

The system has one stable equilibrium at the origin with a bounded region of attraction

$$\mathbf{X}_0 \subset \mathbf{X} := [-1.2, \, 1.2]^2.$$

In order to compute an outer approximation to this region we take $T = 100$ and $\mathbf{X}_T = \{\mathbf{x} \; : \; \|\mathbf{x}\|_2 \leq 0.01\}$. Plots of the ROA estimates $\mathbf{X}_{0d}$ for $2d \in \{10, 12, 14, 16\}$ are shown in Figure 10.1. Observe a relatively-fast convergence of the super-level sets to the ROA — this is confirmed by the relative volume error[3] summarized in Table 10.1. Figure 10.2 then shows the approximating polynomial itself for degree $2d = 18$.

Table 10.1   Van der Pol oscillator — relative error of the outer approximation to the ROA $\mathbf{X}_0$ as a function of the approximating polynomial degree.

degree ($= 2d$)	10	12	14	16
error	49.3 %	19.7 %	11.1 %	5.7 %

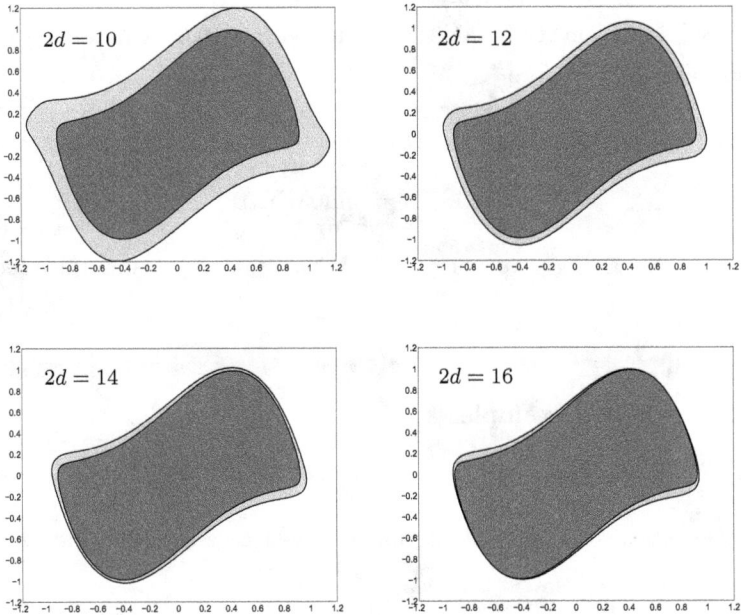

Fig. 10.1   Van der Pol oscillator — semialgebraic outer approximations (light gray) to the ROA (dark gray) for degrees $k \in \{10, 12, 14, 16\}$.

---

[3]The relative volume error was computed approximately by Monte Carlo integration.

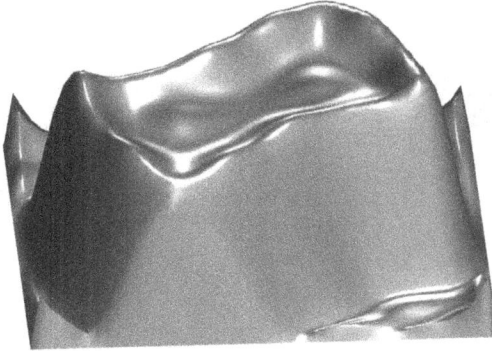

Fig. 10.2   Van der Pol oscillator — a polynomial approximation of degree 18 of the ROA indicator function $I_{\mathbf{X}_0}$.

10.5.2   *Double integrator*

Consider a double integrator $\dot{x}_1 = x_2$; $\dot{x}_2 = u$. The goal is to find an approximation to the set of all initial states $\mathbf{X}_0$ that can be steered to the origin at the final time $T = 1$. To this end, set $X_T = \{0\}$ and the constraint set $\mathbf{X}$ such that $\mathbf{X}_0 \subset \mathbf{X}$, e.g., $\mathbf{X} = [-0.7, 0.7] \times [-1.2, 1.2]$. The solution to this problem can be computed analytically as

$$\mathbf{X}_0 = \{\mathbf{x} \; : \; V(\mathbf{x}) \leq 1\},$$

$$\text{where} \quad V(\mathbf{x}) = \begin{cases} x_2 + 2\sqrt{x_1 + \frac{1}{2}x_2^2} & \text{if } x_1 + \frac{1}{2}x_2|x_2| > 0, \\ -x_2 + 2\sqrt{-x_1 + \frac{1}{2}x_2^2} & \text{otherwise.} \end{cases}$$

The ROA estimates $\mathbf{X}_{0d}$ for $2d \in \{6, 8, 10, 12\}$ are shown in Figure 10.3; again, observe a relatively fast convergence of the super-level set approximations, which is confirmed by the relative volume errors in Table 10.2.

Table 10.2   Double integrator — relative error of the outer approximation to the ROA $\mathbf{X}_0$ as a function of the approximating polynomial degree.

degree ($= 2d$)	6	8	10	12
error	75.7 %	32.6 %	21.2 %	16.0 %

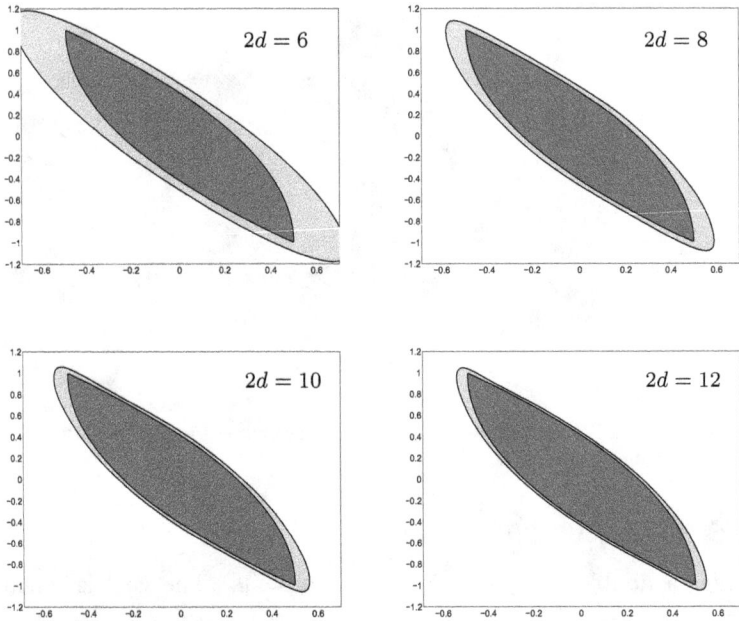

Fig. 10.3   Double integrator — semialgebraic outer approximations (light gray) to the ROA (dark gray) for degrees $2d \in \{6, 8, 10, 12\}$.

10.5.3   *Acrobot*

Consider the acrobot system adapted from Murray & Hauser (1991), which is a double pendulum with both joints actuated; see Figure 10.4. The system equations are given by

$$\dot{\mathbf{x}} = \begin{bmatrix} x_3 \\ x_4 \\ \mathbf{M}(\mathbf{x})^{-1}\mathbf{N}(\mathbf{x},\mathbf{u}) \end{bmatrix} \in \mathbb{R}^4, \quad \text{where} \quad \mathbf{M}(\mathbf{x}) = \begin{bmatrix} 3+\cos(x_2) & 1+\cos(x_2) \\ 1+\cos(x_2) & 1 \end{bmatrix}$$

and $\mathbf{N}(\mathbf{x},\mathbf{u}) = \begin{bmatrix} g\sin(x_1+x_2) - a_1x_3 + a_2\sin(x_1) + x_4\sin(x_2)(2x_3+x_4) + u_1 \\ -\sin(x_2)x_3^2 - a_1x_4 + g\sin(x_1+x_2) + u_2 \end{bmatrix}$,

with $g = 9.8$, $a_1 = 0.1$ and $a_2 = 19.6$. The first two states are the joint angles (in radians) and the second two the corresponding angular velocities (in radians per second). The two control inputs are the torques in the two joints. The goal is to study how the size of the ROA approximations is influenced by the actuation of the first joint. We consider two cases:

with both joints actuated and with only the middle joint actuated. In the
first case the input constraint set is $\mathbf{U} = [-10, 10] \times [-10, 10]$ and in the
second case it is $\mathbf{U} = \{0\} \times [-10, 10]$. The state constraint set is for both
cases $\mathbf{X} = [-\pi/2, \pi/2] \times [-\pi, \pi] \times [-5, 5] \times [-5, 5]$. Since this system is
not polynomial we take a third-order Taylor expansion of the vector field
around the origin. An exact treatment would be possible via a coordinate
transformation leading to rational dynamics to which our approach can
be readily extended. Figure 10.5 shows the approximations $\mathbf{X}_{0d}$ of degree
$2d \in \{6, 8\}$; as expected disabling actuation of the first joint leads to a
smaller ROA approximation. Before solving, the problem data was scaled
such that the constraint sets become unit boxes.

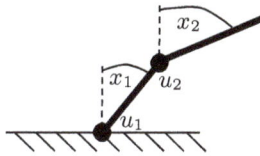

Fig. 10.4   Acrobot — sketch

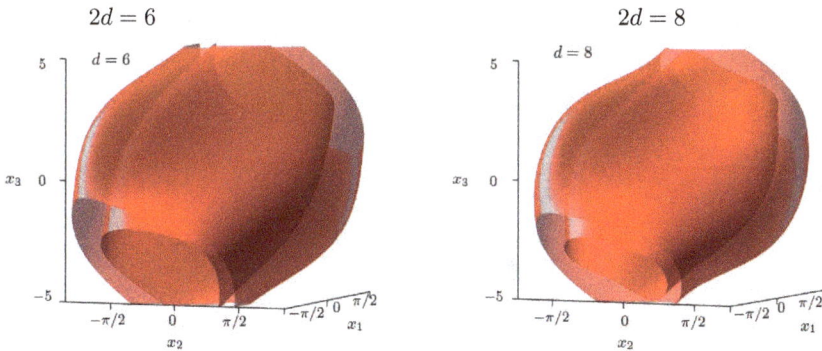

Fig. 10.5   Acrobot — section for $x_4 = 0$ of the semialgebraic outer approximations
of degree $2d \in \{6, 8\}$. Only the middle joint actuated – darker, smaller; both joints
actuated – lighter, larger. The states displayed x_1, x_2 and x_3 are, respectively, the lower
pendulum angle, the upper pendulum angle and the lower pendulum angular velocity.

10.6 Notes and sources

This chapter is based on Henrion & Korda (2014) and Korda (2016); a modification to obtain inner approximations is described in Korda et al. (2013) and the problem of maximum controlled invariant set computation (both in discrete and continuous time) is treated in Korda et al. (2014). Historically, the idea of a linear embedding of a nonlinear dynamical system using occupation measures goes back (at least) to the work Vinter & Lewis (1978), building on the seminal work Young (1969). The subsequent relaxation of the infinite-dimensional LP using semidefinite programming was, to the best of our knowledge, first carried out in Lasserre et al. (2008) in the context of optimal control. The distinctive feature of the presented approach is its overall convexity, completely circumventing the inherent pitfalls of traditional approaches based on non-convex optimization techniques such as initialization, local optima or saddle points; see Henrion & Korda (2014) for a survey of existing approaches for ROA computation. The proof of Theorem 10.2 relies on the superposition principle Ambrosio (2008), Theorem 3.2, and can be found in the Appendix A of Henrion & Korda (2014). Assumption 10.1(b) says that, at any $x \in \mathbf{X}$, the set of all attainable vector fields by admissible control inputs is convex; if this assumption does not hold it may happen that $p^\star > \mathrm{vol}\,\mathbf{X}_0$; see Section III-B and Appendix B of Henrion & Korda (2014) for a discussion and an example.

Bibliography

Ambrosio, L. (2008). *Transport equation and Cauchy problem for non-smooth vector fields.* In Calculus of variations and nonlinear partial differential equations (pp. 1–41). Springer, Berlin, Heidelberg.

Murray, R. M. and Hauser, J. E. (1991). *A case study in approximate linearization: The acrobat example.* Electronics Research Laboratory, College of Engineering, University of California.

Korda, M., Henrion, D. and Jones, C. N. (2014). *Convex computation of the maximum controlled invariant set for polynomial control systems.* SIAM Journal on Control and Optimization, 52(5), 2944–2969.

Young, L. C. (1969). *Lecture on the calculus of variations and optimal control theory.* W. B. Saunders, Philadelphia.

Vinter, R. B. and Richard M. Lewis (1978). *The equivalence of strong and weak formulations for certain problems in optimal control.* SIAM Journal on Control and Optimization 16.4: 546–570.

Henrion, D. and Korda, M. (2014). *Convex computation of the region of attraction of polynomial control systems.* IEEE Transactions on Automatic Control, 59(2), 297–312.

Korda, M., Henrion, D. and Jones, C. N. (2013). Inner approximations of the region of attraction for polynomial dynamical systems. IFAC Proceedings Volumes, 46(23), 534–553.

Korda, M. (2016). *Moment-sum-of-squares hierarchies for set approximation and optimal control.* Doctoral thesis, École polytechnique fédérale de Lausanne.

Lasserre, J. B., Henrion, D., Prieur, C. and Trélat, E. (2008). *Nonlinear optimal control via occupation measures and LMI-relaxations*. SIAM journal on control and optimization, 47(4), 1643–1666.

Bagby, Thomas, Len Bos, and Norman Levenberg (2002). *Multivariate simultaneous approximation*. Constructive approximation 18.4: 569–577.

Anderson, Edward J. and Nash, P. (1987). *Linear Programming in Infinite-Dimensional Spaces: Theory and Applications*, John Wiley and Sons.

Lasserre, J. B. (2009). *Moments, Positive Polynomials and Their Applications* (Imperial College Press, London, UK).

Chapter 11

Non-Linear Partial Differential Equations

We describe how to apply the Moment-SOS hierarchy to solve a class of non-linear partial differential equations (PDE).

11.1 Introduction

In this chapter we show how to use the Moment-SOS hierarchy to solve approximately scalar non-linear hyperbolic conservation laws, a partial differential equation (PDE) which models numerous physical phenomena such as fluid mechanics, traffic flow or nonlinear acoustics Dafermos (2006), Whitham (2011). The existence and uniqueness of solutions to the associated Cauchy problem crucially depends on the flux and the initial condition Kruzkhov (1970). Even if the solution is unique, its numerical computation is still a challenge — in particular when the solution has a shock, i.e., a discontinuity. Existing schemes based on discretization such as Godunov (1959) suffer from numerical dissipation: the shock is smoothened in the numerical solution and cannot be represented accurately. In fact, sometimes the exact location of the shock is of crucial interest for applications. Note however that some existing numerical schemes are able to capture shock in the case where the conservation laws under consideration are linear, see e.g. Després and F. Lagoutière (2001).

In contrast with existing methods, a distinguishing feature of the Moment-SOS hierarchy is to *not* rely on time or space discretization; it computes the solution in a given time-space window globally. From such a solution, the location of the shock at a given time can be computed up to the limits of machine precision. In our opinion this is a major advantage when compared to other numerical methods.

Measure-Valued Solutions. While PDEs are usually understood in a weak sense, DiPerna proposed an even weaker notion of solution, the so-called *measure-valued solutions* (mv solutions for short) DiPerna (1985), which are based on Young measures, or parametrized (i.e. time and/or space dependent) probability measures. Young measures have originally been introduced in the context of calculus of variations and optimal control, where the velocity or more generally the control is relaxed from being a function of time to being a time-dependent probability measure on the control space, see e.g. (Fattorini, 1999, Part III) for an overview. Similarly, DiPerna introduced mv solutions to conservation laws as measures on the solution space, now depending on time and space.

Naturally, every weak solution gives rise to a mv solution when identifying a solution $y(t, \mathbf{x})$ with the Young measure $\delta_{y(t,\mathbf{x})}(dy)$. We say then that the mv solution is *concentrated* on (the graph of) the solution. In this paper, we are focusing on a setup where both weak and mv solutions are unique (hence identical). In this case, both solutions coincide via the identification just mentioned. Note however that our approach also applies without any change to situations where the mv solution is not concentrated, e.g., because of an initial condition that is not concentrated either.

In order to ensure uniqueness we rely on the notion of *entropy solutions* which has been extended to entropy mv solutions. Entropy is a concept from thermodynamics that makes reference to the fact that differences in physical systems, e.g., the densities of particles in a room, tend to adjust to each other. It is well-known that the entropy solution of a scalar non-linear hyperbolic conservation law is unique. For the generalized situation things are more involved. However under suitable assumptions on the initial condition entropy, uniqueness of mv solutions can be proved.

Recently there has been an increasing interest in numerical schemes to compute mv solutions for hyperbolic conservation laws Fjordholm et al. (2017); Feireisl et al. (2019). In comparison, our approach rather computes numerically the *moments* of the unique mv solution and it does not rely on a time-space discretization. In a second step, the time and space dependent solution $(t, \mathbf{x}) \mapsto y(t, \mathbf{x})$ can be recovered from the computed moments of the mv solution.

Generalized Moment Problem. As a common feature of all the results described in this book, the key idea underlying the approach is to consider mv solutions as solutions to a particular instance of the *Generalized Moment Problem* (GMP). Then moments of mv solutions are approximated as closely as desired by solving a specific Moment-SOS hierarchy which follows the general scheme in Figure 2.1. Interestingly, it is worth noting that Fjordholm et al. (2017) already pointed out that the statistical moments of mv solutions are precisely the quantities of interest.

11.2   Notions of solutions

We start with a brief overview of different notions of solutions to scalar polynomial PDEs. For details, we refer to Dafermos (2006) for weak solutions and Malek et al. (1996) for measure-valued solutions. The aim of this section is to give a clear link between these two concepts of solutions.

11.2.1   *Weak and entropy solutions*

In order to study mv solutions, it is instructive to revisit the classical concept of weak solutions first. Consider therefore the Cauchy problem

$$\frac{\partial y}{\partial t}(t, x) + \frac{\partial f(y)}{\partial x}(t, x) = 0, \quad (t, x) \in \mathbb{R}_+ \times \mathbb{R}, \qquad (11.1a)$$

$$y(0, x) = y_0(x), \quad x \in \mathbb{R}, \qquad (11.1b)$$

where (11.1a) is a scalar hyperbolic conservation law with $f \in \mathbb{C}^1(\mathbb{R})$ and (11.1b) provides an initial condition $y_0 \in \mathscr{L}^1(\mathbb{R}) \cap \mathscr{L}^\infty(\mathbb{R})$. Note that (11.1a) encompasses, among others, the well-known Burgers equation if one sets $f(y) = \frac{1}{2}y^2$.

Even if the initial condition y_0 is smooth, solutions to (11.1) might be discontinuous (see (Evans, 2010, p. 143) for the case of the Burgers equation). Solutions to this problem are hence usually understood in the following weak sense.

Definition 11.1 (Weak solution). A function $y \in \mathscr{L}^\infty(\mathbb{R}_+ \times \mathbb{R})$ is a weak solution to (11.1) if, for all test functions $\psi_1 \in \mathscr{C}_c^1(\mathbb{R}_+ \times \mathbb{R})$, it satisfies

$$\int_{\mathbb{R}_+} \int_{\mathbb{R}} \left(\frac{\partial \psi_1(t,x)}{\partial t} y(t,x) + \frac{\partial \psi_1(t,x)}{\partial x} f(y(t,x)) \right) dx\, dt$$

$$+ \int_{\mathbb{R}} \psi_1(0,x) y_0(x)\, dx = 0. \tag{11.2}$$

In general, weak solutions to (11.1) are not unique. However it can be shown (see e.g. Kruzkhov (1970)) that among all possible weak solutions, only one has a physical meaning. This solution is called an *entropy solution* and it can be characterized as follows.

Definition 11.2 (Entropy pair/entropy solution).

(i) A pair of functions $\eta, q \in \mathscr{C}^1(\mathbb{R})$ is called an *entropy pair* for (11.1a) if η is strictly convex and $q' = f'\eta'$.

(ii) A weak solution $y \in \mathscr{L}^\infty(\mathbb{R}_+ \times \mathbb{R})$ of (11.1) is an *entropy solution* if, for all entropy pairs and all non-negative test functions $\psi_2 \in \mathscr{C}_c^1(\mathbb{R}_+ \times \mathbb{R})$, it satisfies

$$\int_{\mathbb{R}_+} \int_{\mathbb{R}} \left(\frac{\partial \psi_2}{\partial t} \eta(y) + \frac{\partial \psi_2}{\partial x} q(y) \right) dx\, dt + \int_{\mathbb{R}} \psi_2(0,x) \eta(y_0)(x)\, dx \geq 0. \tag{11.3}$$

11.2.2 Measure-valued solutions

Generally, regularity results of conservation laws are obtained from regularized conservation laws

$$\frac{\partial}{\partial t} y^\varepsilon + \frac{\partial}{\partial x} f(y^\varepsilon) - \varepsilon \frac{\partial^2}{\partial x^2} y^\varepsilon = 0,$$

where $\varepsilon > 0$ is a fixed parameter. Then one studies the limit of solutions y^ε as ε goes to 0 and tries to retrieve some of regularity properties of the latter equation for the conservation law. However, on the one hand, regularized solution y^ε may or may not converge to a weak solution y of (11.1). This is due of a lack of reflexivity of the space $\mathscr{L}^{\infty}$. On the other hand, regularized solutions y^{ε} necessarily converge to a measure-valued (mv) solution. This notion builds upon the concept of a Young measure.

Definition 11.3 (Young measure). A Young measure on a Euclidean space $\mathcal{X}$ is a map $\mu : \mathcal{X} \to \mathscr{P}(\mathbb{R})$, $\xi \mapsto \mu_{\xi}$, such that for all $g \in \mathscr{C}_0(\mathbb{R})$ the function $\xi \mapsto \int_{\mathbb{R}} g(y)\mu_{\xi}(dy)$ is measurable.

Later, mv solutions have also proved to be useful in the study of problems more general than (11.1), where the initial condition (11.1b) is replaced by a Young measure parametrized in space (see e.g. Fjordholm et al. (2017) and the references therein). The generalized problem is to find a Young measure $\mu_{(t,x)}$ which satisfies the following Cauchy problem:

$$\partial_t \langle \mu_{(t,x)}, y \rangle + \partial_x \langle \mu_{(t,x)}, f(y) \rangle = 0, \quad (t,x) \in \mathbb{R}_+ \times \mathbb{R}, \qquad (11.4a)$$

$$\mu_{(0,x)} = \sigma_0, \quad x \in \mathbb{R}, \qquad (11.4b)$$

where $\langle \cdot, \cdot \rangle$ denotes integration of a measure $\mu \in \mathscr{M}_+(\mathbb{R})$ against a function $g \in \mathscr{C}(\mathbb{R})$:

$$\langle \mu, g \rangle := \int_{\mathbb{R}} g(y)\mu(dy).$$

In (11.4) the measure σ_0 is a given Young measure on $\mathbb{R}$, and f is a continuously differentiable function on $\mathbb{R}$. The conservation law (11.4a) has to be understood in the sense of distributions, i.e.:

Definition 11.4 (Measure-valued solution). A Young measure is a measure-valued (mv) solution to (11.4) if, for all test functions $\psi_1 \in \mathscr{C}_c^1(\mathbb{R}_+ \times \mathbb{R})$, it satisfies

$$\int_{\mathbb{R}_+} \int_{\mathbb{R}} \left(\frac{\partial \psi_1(t,x)}{\partial t} \langle \mu_{(t,x)}, y \rangle + \frac{\partial \psi_1(t,x)}{\partial x} \langle \mu_{(t,x)}, f(y) \rangle \right) dx \, dt$$

$$+ \int_{\mathbb{R}} \psi_1(0,x) \langle \sigma_0, y \rangle dx = 0. \qquad (11.5)$$

Note that the weak solution y has been replaced by a time-space parametrized probability measure μ supported on the range of y. Whereas

a weak solution is requested to satisfy (11.2), only averages of the mv solution are considered in (11.5). It is easy to see that every weak solution
induces a mv solution via the canonical embedding $y(t,x) \mapsto \delta_{y(t,x)}$. As in
the case of weak solution, an entropy condition is needed in order to select
solutions with a physical meaning. Quite in analogy to entropy solutions,
entropy mv solutions are defined as follows.

Definition 11.5 (Entropy measure-valued solution). An mv solution μ is an entropy mv solution to (11.4) if, for all entropy pairs (η, q)
and all non-negative test functions $\psi_2 \in \mathscr{C}_c^1(\mathbb{R}_+ \times \mathbb{R})$, it satisfies

$$\int_{\mathbb{R}_+} \int_{\mathbb{R}} \left(\frac{\partial \psi_2(t,x)}{\partial t} \langle \mu_{(t,x)}, \eta \rangle + \frac{\partial \psi_2(t,x)}{\partial x} \langle \mu_{(t,x)}, q \rangle \right) dx\, dt$$

$$+ \int_{\mathbb{R}} \psi_2(0,x) \langle \sigma_0, \eta \rangle dx \geq 0. \tag{11.6}$$

Remark 11.1. Again it is straightforward to see that entropy solutions
are entropy mv solutions via the canonical embedding $y(t,x) \mapsto \delta_{y(t,x)}$.
However, as demonstrated on an example in (Fjordholm et al., 2017, p. 775),
in contrast with entropy solutions, entropy mv solutions are not necessarily
unique.

We have seen that the concept of mv solutions is weaker than the concept
of weak solutions. Hence mv solutions are a relaxation of weak solutions:
every weak solution is also an mv solution, but the set of mv solutions can
be larger than the set of weak solutions. However, the following result states
that considering mv solutions is not a relaxation. To be more precise, when
the initial measure in (11.4b) is concentrated on (the graph of) the initial
condition in (11.1b), then the entropy mv solution to (11.4) is unique and
concentrated on (the graph of) the (unique) entropy solution to (11.1).

Theorem 11.2 (Concentration of the entropy mv solution). *Let C
be the Lipschitz constant of the function f. Let y be an entropy solution
and μ be an entropy mv solution to (11.1). Then, for all $T \geq 0$ and all*

$r \geq 0$, *it holds*

$$\int_{|x|\leq r} \langle \mu_{(t,x)}, |y - y(T,x)|\rangle dx \leq \int_{|x|\leq r+CT} \langle \sigma_0, |y - y_0(x)|\rangle dx. \quad (11.7)$$

In particular, if $\sigma_0 = \delta_{y_0(x)}$, then $\mu_{(t,x)} = \delta_{y(t,x)}$ for all $t \in [0,T]$ and all x such that $|x| \leq r$.

Remark 11.3. The proof of Theorem 11.2 is similar to the one provided in Kruzkhov (1970) and it can be found in the Appendix of Marx et al. (2019). It is based on the doubling variable strategy, using the following family of entropy pairs:

$$\eta_v(y) = |y - v|, \quad q_v(y) = \text{sign}(y - v)(f(y) - f(v)) \quad (11.8)$$

parametrized in $v \in \mathbb{R}$. In Lax (1971), it has been proved that a linear combination of these entropy pairs, together with the convex hull of linear functions, generate all entropy pairs. In other words, to prove Theorem 11.2 for every entropy pair it is enough to consider the entropy pairs (11.8).

Moreover, note that initially the doubling variable strategy has been used to prove uniqueness of the solution to scalar non-linear conservations laws. The main drawback is that the entropy solution has to satisfy this inequality for all convex pairs. However for the specific case of the Burgers equation, it is shown in DeLellis et al. (2004) and Panov (1994) that one may consider only one convex pair.

11.2.3 *An emphasis on compact sets*

In practice, one computes or approximates the solution on compact subsets, so let

$$\mathbf{T} := [0,T], \quad \mathbf{X} := [L,R] \quad (11.9)$$

be the respective domains of time t and space x, for fixed (but arbitrary) constants T, L, R. After scaling, we assume without loss of generality that $T = R - L = 1$.

Note that the entropy inequality induces a stability property:

$$\|y(t,\cdot)\|_{\mathscr{L}^\infty(\mathbf{X})} \leq \|y_0\|_{\mathscr{L}^\infty(\mathbb{R})}, \quad \forall t \geq 0 \quad (11.10)$$

see e.g. (Dafermos, 2006, Theorem 6.2.4). Since y_0 is bounded in $\mathscr{L}^\infty$, it follows from the maximum principle (Dafermos, 2006, Theorem 6.3.2) that $y(t, .)$ is bounded in $\mathscr{L}^\infty$ for all $t \geq 0$. Hence, we can consider that y takes values in the following compact set

$$\mathbf{Y} := [\underline{y}, \bar{y}], \tag{11.11}$$

where the bounds $\underline{y} := \text{ess inf}_{x \in \mathbb{R}}\, y_0(x)$ and $\bar{y} := \text{ess sup}_{x \in \mathbb{R}}\, y_0(x)$ depend on the initial condition. On $\mathbf{T} \times \mathbf{X}$, the polynomial hyperbolic equations given in (11.1) reads:

$$\begin{cases} \dfrac{\partial y}{\partial t} + \dfrac{\partial f(y)}{\partial x} = 0, & (t, x) \in \mathbf{T} \times \mathbf{X}, \\[2mm] y(0, x) = y_0(x), & x \in \mathbf{X}. \end{cases} \tag{11.12}$$

Definition 11.6 (Entropy solution on compact sets). A weak solution y is an entropy solution to (11.12) if, for all test functions $\psi_1 \in \mathscr{C}^1(\mathbb{R}_+ \times \mathbb{R})$, it satisfies

$$\int_\mathbf{T} \int_\mathbf{X} \left(\frac{\partial \psi_1}{\partial t} y + \frac{\partial \psi_1}{\partial x} f(y) \right) dx dt + \int_\mathbf{X} \psi_1(0, x) y_0(x) dx$$
$$- \int_\mathbf{X} \psi_1(T, x) y(T, x) dx + \int_\mathbf{T} \psi_1(t, L) y(t, L) dt - \int_\mathbf{T} \psi_1(t, R) y(t, R) dt = 0 \tag{11.13}$$

and, for all convex pairs (p, q) and all non-negative test functions $\psi_2 \in \mathscr{C}^1(\mathbb{R}_+ \times \mathbb{R})$, it satisfies

$$\int_\mathbf{T} \int_\mathbf{X} \left(\frac{\partial \psi_2}{\partial t} \eta(y) + \frac{\partial \psi_2}{\partial x} q(y) \right) dx dt + \int_\mathbf{X} \psi_2(0, x) \eta(y_0)(x) dx$$
$$- \int_\mathbf{X} \psi_2(T, x) \eta(y(T, x)) dx + \int_\mathbf{T} \psi_2(t, L) q(y(t, L)) dt$$
$$- \int_\mathbf{T} \psi_2(t, R) q(y(t, R)) dt \geq 0. \tag{11.14}$$

As we work on compact sets, the test functions do not have to vanish at infinity. However new terms $y(t, R)$ and $y(t, L)$ now appear. Related to this notion of solutions on compact sets, we also have a similar definition for measure-valued (mv) entropy solution.

Definition 11.7 (mv entropy solution on compact sets). A Young measure $\mu : (t,x) \in \mathbf{T} \times \mathbf{X} \mapsto \mu_{(t,x)} \in \mathscr{P}(\mathbf{Y})$ is an entropy measure-valued solution to (11.12) if, for all test functions $\psi_1 \in \mathscr{C}^1(\mathbf{T} \times \mathbf{X})$,

$$\int_{\mathbf{T}} \int_{\mathbf{X}} \left(\frac{\partial \psi_1}{\partial t} \langle \mu_{(t,x)}, y \rangle + \frac{\partial \psi_1}{\partial x} \langle \mu_{(t,x)}, f(y) \rangle \right) dx\, dt$$

$$+ \int_{\mathbf{X}} \psi_1(0,x) \langle \sigma_0, y \rangle dx - \int_{\mathbf{X}} \psi_1(T,x) \langle \sigma_T, y \rangle dx$$

$$+ \int_{\mathbf{T}} \psi_1(t,L) \langle \sigma_L, f(y) \rangle dt - \int_{\mathbf{T}} \psi_1(t,R) \langle \sigma_R, f(y) \rangle dt = 0 \quad (11.15)$$

and, for all convex pairs (η, q) and all non-negative test functions $\psi_2 \in \mathscr{C}^1(\mathbf{T} \times \mathbf{X})$,

$$\int_{\mathbf{T}} \int_{\mathbf{X}} \left(\frac{\partial \psi_2}{\partial t} \langle \mu_{(t,x)}, \eta(y) \rangle + \frac{\partial \psi_2}{\partial x} \langle \mu_{(t,x)}, q(y) \rangle \right) dx\, dt$$

$$+ \int_{\mathbf{X}} \psi_2(0,x) \langle \sigma_0, \eta(y) \rangle dx - \int_{\mathbf{X}} \psi_2(T,x) \langle \sigma_T, \eta(y) \rangle dx$$

$$+ \int_{\mathbf{T}} \psi_2(t,L) \langle \sigma_L, q(y) \rangle dt - \int_{\mathbf{T}} \psi_2(t,R) \langle \sigma_R, q(y) \rangle dt \geq 0, \quad (11.16)$$

where σ_0, σ_T, σ_L, resp. σ_R are Young measures supported on $\mathbf{T}$, $\mathbf{X}$, $\mathbf{T}$, resp. $\mathbf{X}$.

Remark 11.4 (Imposing constraints on the boundary). To ensure the concentration of $\mu_{(t,x)}$ on the graph of the solution to (11.13)-(11.14), in addition to the condition $\sigma_0 = \delta_{y_0(x)}$, one may impose conditions on the boundary measures σ_L and/or σ_R. In practice, one knows the initial condition in an interval larger than $\mathbf{X}$ and so one is able to impose σ_L and/or σ_R. The width of this interval depends on the Lipschitz constant of the flux, T, L and R. As an illustrative example, consider the case where the initial condition is positive and the flux is strictly convex. By the classical method of characteristics, if the initial condition y_0 is positive then so is the solution y for all $t \geq 0$. In particular if f is strictly convex we only need to impose knowledge at the left of the box $\mathbf{X}$. Therefore σ_L has to be known for all $t \in \mathbf{T}$, and σ_R is unconstrained. We refer to

Leveque (1992) for a more precise discussion on the choice of the boundary
constraint.

11.3 A convex optimization approach for mv solutions on compact sets

In the latter section, we introduced mv solutions for scalar hyperbolic equations. Note that measures are fully characterized by their moments on compact sets, see e.g. (Lasserre, 2010, p. 52). This means in particular that
moments are the quantities of interest. The aim of this section is to express
formulations (11.15)-(11.16) as constraints on the moments, to explain how
one can compute numerically these moments thanks to the moment-SOS
hierarchy. We also show how one can interpret these moments in the case
where the initial measure in concentrated.

11.3.1 *Moment constraints for the entropy mv solution*

Let $\nu \in \mathscr{M}(\mathbf{K})_+$, with $\mathbf{K} := \mathbf{T} \times \mathbf{X} \times \mathbf{Y}$. In the following, we derive moment
constraints that will imply that ν can be disintegrated as follows

$$d\nu(t,x,y) = dt\, dx\, d\mu_{(t,x)}(dy) \qquad (11.17)$$

or, equivalently,

$$\nu = \lambda_{\mathbf{T}}\lambda_{\mathbf{X}}\mu_{(t,x)}, \qquad (11.18)$$

where μ is an entropy mv solution satisfying (11.15) and (11.16). In (11.17)
the measure ν is called an *occupation measure* and the Young measure μ is
its conditional measuring y given t and x. We also need to introduce the
following time boundary measures

$$d\nu_0(t,x,y) := \delta_0(dt)\, dx\, \sigma_0(dy), \quad d\nu_T(t,x,y) := \delta_T(dt)\, dx\, \sigma_T(dy) \quad (11.19)$$

whose supports are $\mathbf{K}_0 := \{0\} \times \mathbf{X} \times \mathbf{Y}$ and $\mathbf{K}_T := \{T\} \times \mathbf{X} \times \mathbf{Y}$ respectively.
Similarly, we introduce the following space boundary measures.

$$d\nu_L(t,x,y) := dt\, \delta_L(dx)\, \sigma_L(dy); \quad d\nu_R(t,x,y) := dt\, \delta_R(dx)\, \sigma_R(dy)$$

whose supports are given by $\mathbf{K}_L := \mathbf{T} \times \{L\} \times \mathbf{Y}$ and $\mathbf{K}_R := \mathbf{T} \times \{R\} \times \mathbf{Y}$ respectively.

First, to ensure that the marginal of ν with respect to t and x is the Lebesgue measure on $\mathbf{T} \times \mathbf{X}$, it suffices to impose that:

$$\int_{\mathbf{K}} t^{\alpha_1} x^{\alpha_2} \, d\nu(t, x, y) = \int_{\mathbf{T} \times \mathbf{X}} t^{\alpha_1} x^{\alpha_2} \, dt \, dx, \quad \alpha \in \mathbb{N}^2. \qquad (11.20)$$

In a similar manner, we can enforce the respective marginal of the boundary measures to be Lebesgue as follows

$$\int_{\mathbf{K}_0} 0^{\alpha_1} x^{\alpha_2} \, d\nu_0(t, x, y) = \int_{\mathbf{X}} 0^{\alpha_1} x^{\alpha_2} \, dx, \quad \alpha \in \mathbb{N}^2, \qquad (11.21)$$

$$\int_{\mathbf{K}_T} T^{\alpha_1} x^{\alpha_2} \, d\nu_T(t, x, y) = \int_{\mathbf{X}} T^{\alpha_1} x^{\alpha_2} \, dx, \quad \alpha \in \mathbb{N}^2, \qquad (11.22)$$

$$\int_{\mathbf{K}_L} t^{\alpha_1} L^{\alpha_2} \, d\nu_L(t, x, y) = \int_{\mathbf{T}} t^{\alpha_1} L^{\alpha_2} \, dt, \quad \alpha \in \mathbb{N}^2 \qquad (11.23)$$

and

$$\int_{\mathbf{K}_R} t^{\alpha_1} R^{\alpha_2} \, d\nu_R(t, x, y) = \int_{\mathbf{T}} t^{\alpha_1} R^{\alpha_2} \, dt, \quad \alpha \in \mathbb{N}^2. \qquad (11.24)$$

Next, we aim at proving that (11.15) and (11.16) can also be expressed by moment constraints. We split the exposition into two steps: the first one deals with (11.15), while the second one deals with (11.16).

11.3.1.1 *First step: enforcing (11.15) by moment constraints*

Lemma 11.5. *Let* $\phi_1^\alpha(t, x, y) := t^{\alpha_1} x^{\alpha_2} y$ *and* $\phi_2^\alpha(t, x, y) := t^{\alpha_1} x^{\alpha_2} f(y)$ *for* $\alpha \in \mathbb{N}^2$. *Linear constraint (11.15) is equivalent to*

$$\int_{\mathbf{K}} \left(\frac{\partial \phi_1^\alpha}{\partial t} y + \frac{\partial \phi_2^\alpha}{\partial x} f(y) \right) d\nu + \int_{\mathbf{K}_0} \phi_1^\alpha \, d\nu_0 - \int_{\mathbf{K}_T} \phi_1^\alpha \, d\nu_T$$
$$+ \int_{\mathbf{K}_L} \phi_2^\alpha \, d\nu_L - \int_{\mathbf{K}_R} \phi_2^\alpha \, d\nu_R = 0 \qquad (11.25)$$

for all $\alpha \in \mathbb{N}^2$.

Proof. Since $\mathbf{T} \times \mathbf{X}$ is a compact set, as a consequence of the Stone-Weierstrass theorem, we can restrict the test functions to $\psi_1 = t^{\alpha_1} x^{\alpha_2}$ for $\alpha \in \mathbb{N}^2$ to enforce (11.15). $\qquad\qquad\square$

11.3.1.2 *Second step: enforcing (11.16) by moment constraints*

As noticed in Remark 11.3, the entropy inequality is satisfied for all the convex pairs (η, q) if and only it is satisfied for all Kruzkhov entropies given in (11.8). To express (11.16) as moment constraints, we are faced with two issues: first, taking into account an uncountable family of functions parametrized by $v \in \mathbf{Y}$ and, second, the absolute value function $v \mapsto |v|$ is not a polynomial. To deal with the uncountable family of functions, we introduce v as a new variable. To treat the absolute value, we double the number of measures.

More precisely, we define the Borel measures ϑ^+, ϑ^- whose supports are defined as follows

$$\mathbf{K}^+ := \mathrm{spt}\vartheta^+ = \{(t, x, y, v) \in \mathbf{K} \times \mathbf{Y} : y \geq v\},$$

$$\mathbf{K}^- := \mathrm{spt}\vartheta^- = \{(t, x, y, v) \in \mathbf{K} \times \mathbf{Y} : y \leq v\}.$$

Similarly, we define the time boundary measures ϑ_0^+, ϑ_0^-, ϑ_T^+ and ϑ_T^- with the following supports

$$\mathbf{K}_0^+ := \mathrm{spt}\vartheta_0^+ = \{(t, x, y, v) \in \mathbf{K}_0 \times \mathbf{Y} : y \geq v\},$$
$$\mathbf{K}_0^- := \mathrm{spt}\vartheta_0^- = \{(t, x, y, v) \in \mathbf{K}_0 \times \mathbf{Y} : y \leq v\}, \tag{11.26}$$

and

$$\mathbf{K}_T^+ := \mathrm{spt}\vartheta_T^+ = \{(t, x, y, v) \in \mathbf{K}_T \times \mathbf{Y} : y \geq v\},$$
$$\mathbf{K}_T^- := \mathrm{spt}\vartheta_T^- = \{(t, x, y, v) \in \mathbf{K}_T \times \mathbf{Y} : y \leq v\}. \tag{11.27}$$

Finally, let us define the space boundary measures ϑ_L^+, ϑ_L^-, ϑ_R^+ and ϑ_R^- with the following supports

$$\mathbf{K}_L^+ := \mathrm{spt}\vartheta_L^+ = \{(t, x, y, v) \in \mathbf{K}_L \times \mathbf{Y} : y \geq v\}$$
$$\mathbf{K}_L^- := \mathrm{spt}\vartheta_L^- = \{(t, x, y, v) \in \mathbf{K}_L \times \mathbf{Y} : y \leq v\} \tag{11.28}$$

and

$$\mathbf{K}_R^+ := \operatorname{spt}\vartheta_R^+ = \{(t,x,y,v) \in \mathbf{K}_R \times \mathbf{Y} : y \geq v\},$$
$$\mathbf{K}_R^- := \operatorname{spt}\vartheta_R^- = \{(t,x,y,v) \in \mathbf{K}_R \times \mathbf{Y} : y \leq v\}. \tag{11.29}$$

We are now in position to state the following lemma.

Lemma 11.6 (Recovering all Kruzkhov entropies). *Assume that*

$$\vartheta^+ + \vartheta^- = \nu \otimes \lambda_{\mathbf{Y}}, \tag{11.30}$$

$$\vartheta_0^+ + \vartheta_T^- = \nu_0 \otimes \lambda_{\mathbf{Y}}, \quad \vartheta_T^+ + \vartheta_T^- = \nu_T \otimes \lambda_{\mathbf{Y}}, \tag{11.31}$$

$$\vartheta_L^+ + \vartheta_L^- = \nu_L \otimes \lambda_{\mathbf{Y}}, \quad \vartheta_R^+ + \vartheta_R^- = \nu_R \otimes \lambda_{\mathbf{Y}}. \tag{11.32}$$

Then, (11.16) is equivalent to

$$\int_{\mathbf{K}^+} \theta(v) \left(\frac{\partial \psi_2}{\partial t}(y-v) + \frac{\partial \psi_2}{\partial x}(f(y) - f(v)) \right) d\vartheta^+$$
$$+ \int_{\mathbf{K}^-} \theta(v) \left(\frac{\partial \psi_2}{\partial t}(v-y) + \frac{\partial \psi_2}{\partial x}(f(v) - f(y)) \right) d\vartheta^-$$
$$+ \int_{\mathbf{K}_0^+} \theta(v)\psi_2(0,x)(y-v)d\vartheta_0^+ + \int_{\mathbf{K}_0^-} \theta(v)\psi_2(0,x)(v-y) \, d\vartheta_0^-$$
$$- \int_{\mathbf{K}_T^+} \theta(v)\psi_2(T,x)(y-v)d\vartheta_T^+ - \int_{\mathbf{K}_T^-} \theta(v)\psi_2(T,x)(v-y) \, d\vartheta_T^-$$
$$+ \int_{\mathbf{K}_L^+} \theta(v)\psi_2(t,L)(f(y) - f(v))d\vartheta_L^+ + \int_{\mathbf{K}_L^-} \theta(v)\psi_2(t,L)(f(v) - f(y)) \, d\vartheta_L^-$$
$$- \int_{\mathbf{K}_R^+} \theta(v)\psi_2(t,R)(f(y) - f(v))d\vartheta_R^+ - \int_{\mathbf{K}_R^-} \theta(v)\psi_2(t,R)(f(v) - f(y)) \, d\vartheta_R^-$$
$$\geq 0,$$
$$\tag{11.33}$$

for all nonnegative functions $\theta \in \mathbb{C}(\mathbf{Y})$.

Note that from the Stone-Weierstrass Theorem, the constraints (11.30), (11.31) and (11.32) can be expressed as moment constraints: (11.30) holds if and only if, for all $\alpha \in \mathbb{N}^4$,

$$
\int_{\mathbf{K}^+ \cup \mathbf{K}^-} t^{\alpha_1} x^{\alpha_2} y^{\alpha_3} v^{\alpha_4} d(\vartheta^+ + \vartheta^-)(t, x, y, v)
$$
$$
= \int_{\mathbf{K}} t^{\alpha_1} x^{\alpha_2} y^{\alpha_3} d\nu(t, x, y) \int_{\mathbf{Y}} v^{\alpha_4} dv \qquad (11.34)
$$

and similarly for (11.31) and (11.32).

Proof. For conciseness, we deal only with the constraint (11.30). The other constraints of Lemma 11.6 follow similarly. Let us prove that if for all nonnegative functions $\theta \in \mathbb{C}^1(\mathbf{Y})$ and all nonnegative functions $\psi_2 \in \mathbb{C}^1(\mathbf{T} \times \mathbf{X})$,

$$
\int_{\mathbf{K}^+} \theta(v) \left(\frac{\partial \psi_2}{\partial t}(y - v) + \frac{\partial \psi_2}{\partial x}(f(y) - f(v)) \right) d\vartheta^+ \qquad (11.35)
$$
$$
+ \int_{\mathbf{K}^-} \theta(v) \left(\frac{\partial \psi_2}{\partial t}(v - y) + \frac{\partial \psi_2}{\partial x}(f(v) - f(y)) \right) d\vartheta^- =
$$
$$
\int_{\mathbf{K}^+ \cup \mathbf{K}^-} \theta(v) \left(\frac{\partial \psi_2}{\partial t}|y - v| + \frac{\partial \psi_2}{\partial x}\text{sign}(y - v)(f(y) - f(v)) \right) d(\vartheta^+ + \vartheta^-) \geq 0
$$

then the following inequality

$$
\int_{\mathbf{T} \times \mathbf{X} \times \mathbf{Y}} \frac{\partial \psi_2}{\partial t}|y - v| + \frac{\partial \psi_2}{\partial x}\text{sign}(y - v)(f(y) - f(v)) \, dv \geq 0, \qquad (11.36)
$$

holds, for all test functions $\psi_2 \in \mathbb{C}^1(\mathbf{T} \times \mathbf{X})$ and all $v \in \mathbf{Y}$.

First, observe that (11.30) implies that

$$
\int_{\mathbf{K}^+ \cup \mathbf{K}^-} \theta(v) \left(\frac{\partial \psi_2}{\partial t}|y - v| + \frac{\partial \psi_2}{\partial x}\text{sign}(y - v)(f(y) - f(v)) \right) d(\vartheta^+ + \vartheta^-) =
$$
$$
\int_{\mathbf{Y}} \theta(v) \left(\int_{\mathbf{T} \times \mathbf{X} \times \mathbf{Y}} \left(\frac{\partial \psi_2}{\partial t}|y - v| + \frac{\partial \psi_2}{\partial x}\text{sign}(y - v)(f(y) - f(v)) \right) dv \right) dv.
$$
$$
\qquad (11.37)
$$

Then, since (11.35) holds for any nonnegative functions θ, and $y - v = |y - v|$ on $\text{spt}\vartheta^+$ (resp. $v - y = |y - v|$ on $\text{spt}\vartheta^-$),

$$
\int_{\mathbf{T} \times \mathbf{X} \times \mathbf{Y}} \left(\frac{\partial \psi_2}{\partial t}|y - v| + \frac{\partial \psi_2}{\partial x}\text{sign}(y - v)(f(y) - f(v)) \right) dv \geq 0. \qquad (11.38)
$$
$\qquad \qquad \qquad \qquad \qquad \qquad \qquad \qquad \qquad \qquad \qquad \qquad \qquad \qquad \square$

Note that from the Stone-Weierstrass Theorem, the constraints (11.30), (11.31) and (11.32) can be expressed as moment constraints: (11.30) holds if and only if, for all $\alpha \in \mathbb{N}^4$,

$$\int_{\mathbf{K}^+ \cup \mathbf{K}^-} t^{\alpha_1} x^{\alpha_2} y^{\alpha_3} v^{\alpha_4} d(\vartheta^+ + \vartheta^-)(t, x, y, v)$$
$$= \int_{\mathbf{K}} t^{\alpha_1} x^{\alpha_2} y^{\alpha_3} d\nu(t, x, y) \int_{\mathbf{Y}} v^{\alpha_4} dv \qquad (11.39)$$

and similarly for (11.31) and (11.32).

In order to express (11.16) as moment constraints, it remains to prove that the functions ψ_2 and θ can be replaced by suitable polynomials. Here, in contrast with the first step where the functions ψ_1 were unconstrained, the functions ψ_2 and θ have to be nonnegative. To address this issue, we again use positivity certificates from real algebraic geometry.

Lemma 11.7. *Let*

$$\phi_1^{+,\alpha}(t, x, y, v) := t^{\alpha_1} (T - t)^{\alpha_2} (x - L)^{\alpha_3} (R - x)^{\alpha_4} (v - \underline{y})^{\alpha_5} (\bar{y} - v)^{\alpha_6} (y - v),$$

$$\phi_1^{-,\alpha}(t, x, y, v) := t^{\alpha_1} (T - t)^{\alpha_2} (x - L)^{\alpha_3} (R - x)^{\alpha_4} (v - \underline{y})^{\alpha_5} (\bar{y} - v)^{\alpha_6} (v - y),$$

$$\phi_2^{+,\alpha}(t, x, y, v) :=$$
$$\quad t^{\alpha_1} (T - t)^{\alpha_2} (x - L)^{\alpha_3} (R - x)^{\alpha_4} (v - \underline{y})^{\alpha_5} (\bar{y} - v)^{\alpha_6} (f(y) - f(v)),$$

$$\phi_2^{-,\alpha}(t, x, y, v) :=$$
$$\quad t^{\alpha_1} (T - t)^{\alpha_2} (x - L)^{\alpha_3} (R - x)^{\alpha_4} (v - \underline{y})^{\alpha_5} (\bar{y} - v)^{\alpha_6} (f(v) - f(y))$$
$$\qquad (11.40)$$

for $\alpha \in \mathbb{N}^6$. Then, (11.16) is equivalent to

$$\int_{\mathbf{K}^+} \left(\frac{\partial \phi_1^{+,\alpha}}{\partial t} + \frac{\partial \phi_2^{+,\alpha}}{\partial x} \right) d\vartheta^+ + \int_{\mathbf{K}^-} \left(\frac{\partial \phi_1^{-,\alpha}}{\partial t} + \frac{\partial \phi_2^{-,\alpha}}{\partial x} \right) d\vartheta^-$$

$$+ \int_{\mathbf{K}_0^+} \phi_1^{+,\alpha} d\vartheta_0^+ + \int_{\mathbf{K}_0^-} \phi_1^{-,\alpha} d\vartheta_0^- - \int_{\mathbf{K}_T^+} \phi_1^{+,\alpha} d\vartheta_T^+ - \int_{\mathbf{K}_T^-} \phi_1^{-,\alpha} d\hat{\vartheta}_T^-$$

$$+ \int_{\mathbf{K}_L^+} \phi_2^{+,\alpha} d\vartheta_L^+ + \int_{\mathbf{K}_L^-} \phi_2^{-,\alpha} d\vartheta_L^- - \int_{\mathbf{K}_R^+} \phi_2^{+,\alpha} d\vartheta_R^+ - \int_{\mathbf{K}_R^-} \phi_2^{-,\alpha} d\vartheta_R^- \geq 0$$
$$\qquad (11.41)$$

for all $\alpha \in \mathbb{N}^6$.

Proof. The proof relies on a result of real algebraic geometry. Again, invoking the Stone-Weierstrass Theorem, in (11.16) we can restrict the test functions ψ_2 and θ to be polynomials. To enforce their positivity, we use Handelman's Positivstellensatz Handelamn (1988) that implies that

$$
\begin{aligned}
\psi_2(t,x) &= \sum_{\alpha_1,\alpha_2,\alpha_3,\alpha_4 \in \mathbb{N}^4} c_\alpha^{\psi_2} t^{\alpha_1}(T-t)^{\alpha_2}(x-L)^{\alpha_3}(R-x)^{\alpha_4}, \\
\theta(v) &= \sum_{\alpha_5,\alpha_5 \in \mathbb{N}} c_\alpha^{\theta}(v-\underline{y})^{\alpha_5}(\bar{y}-v)^{\alpha_6}
\end{aligned}
\tag{11.42}
$$

with finitely many positive real coefficients $c_\alpha^{\psi_2}, c_\alpha^{\theta}$. Now, (11.41) implies that

$$
\begin{aligned}
\sum_{\alpha \in \mathbb{N}^6} c_\alpha^{\psi_2} c_\alpha^{\theta} &\Bigg\{ \int_{\mathbf{K}^+}\left(\frac{\partial \phi_1^{+,\alpha}}{\partial t}+\frac{\partial \phi_2^{+,\alpha}}{\partial x}\right)d\vartheta^+ + \int_{\mathbf{K}^-}\left(\frac{\partial \phi_1^{-,\alpha}}{\partial t}+\frac{\partial \phi_2^{-,\alpha}}{\partial x}\right)d\vartheta^- \\
&+ \int_{\mathbf{K}_0^+}\phi_1^{+,\alpha}d\vartheta_0^+ + \int_{\mathbf{K}_0^-}\phi_1^{-,\alpha}d\vartheta_0^- - \int_{\mathbf{K}_T^+}\phi_1^{+,\alpha}d\vartheta_T^+ - \int_{\mathbf{K}_T^-}\phi_1^{-,\alpha}d\vartheta_T^- \\
&+ \int_{\mathbf{K}_L^+}\phi_2^{+,\alpha}d\vartheta_L^+ + \int_{\mathbf{K}_L^-}\phi_2^{-,\alpha}d\vartheta_L^- - \int_{\mathbf{K}_R^+}\phi_2^{+,\alpha}d\vartheta_R^+ - \int_{\mathbf{K}_R^-}\phi_2^{-,\alpha}d\vartheta_R^- \Bigg\} \geq 0,
\end{aligned}
\tag{11.43}
$$

which, by linearity of the integrals and the derivatives, recovers (11.33) for ψ_2 and θ given in (11.42). Consequently, by Lemma 11.6, (11.33) implies that (11.16) holds.   □

11.3.2   *The Generalized Moment Problem*

The entropy mv problem can be formulated as a Generalized Moment Problem (GMP) where the unknown are moments of finitely many measures. The measures under consideration are $\nu, \nu_T, \nu_0, \nu_R, \nu_L$ and all the measures we have introduced when transforming the Kruzkhov inequality into moment constraints. The supports of the measures correspond to $\mathbf{T}, \mathbf{X}$ and $\mathbf{Y}$. Finally, the polynomials used for integrating w.r.t. the measures are given in (11.25) (conservation law) (11.41) (entropy inequality), (11.39) (Kruzkhov entropies), and (11.20)-(11.24) (boundary measures).

We may also define an objective functional

$$\int_{\mathbf{K}_L} h d\nu + \int_{\mathbf{K}_0} h_0 d\nu_0 + \int_{\mathbf{K}_T} h_T d\nu_T + \int_{\mathbf{K}_L} h_L d\nu_L + \int_{\mathbf{K}_R} h_R d\nu_R, \quad (11.44)$$

with $h, h_0, h_T, h_L, h_R \in \mathbb{R}[t, x, y]$.

If $\sigma_0 = \delta_{y_0(x)}$ with y_0 an initial condition in (11.13)-(11.14) and, in addition, if one imposes suitable boundary measures as exposed in Remark 11.4, then this objective functional is not especially useful to recover the entropy mv solution of scalar hyperbolic PDE, since the corresponding Young measure is concentrated as a consequence of Theorem 11.2: there is nothing to be optimized. However, with such an objective functional, one can compute quantities of interest such as the energy of the solution. Moreover, our aim is to *relax* the GMP in order to solve it numerically and, then, this objective functional might be helpful to accelerate the convergence of the corresponding relaxations. We refer to Section 11.4 for more discussions about the choice of objective functionals for the Riemann problem of the Burgers equation.

Finally, one is able to define a GMP describing entropy mv solution:

$$\inf_{\nu, \nu_T, \nu_L \text{ and/or } \nu_R}$$ (11.44) (objective functional)

s.t.   (11.25) (conservation law),
       (11.41) (entropy inequality),
       (11.39) (Kruzkhov entropies),
       (11.20)-(11.24) (boundary measures),
       $\nu \in \mathcal{M}(\mathbf{K})_+$ (occupation measure),
       $\nu_T \in \mathcal{M}(\mathbf{K}_T)_+$ (time boundary measure),
       $\nu_L \in \mathcal{M}(\mathbf{K}_L)_+$ (space boundary measure),
       and/or $\nu_R \in \mathcal{M}(\mathbf{K}_R)_+$ (space boundary measures)
$$(11.45)$$
where the measures defined in (11.30)-(11.32) and related to the Kruzkhov entropies are considered as implicit variables.

Then we can apply the moment-SOS hierarchy to solve this GMP numerically by following the general strategy described in Figure 2.1.

11.3.3 Interpretation of the moment solutions

An optimal solution at a given step of the moment-SOS hierarchy of relaxations adapted to the GMP (11.45), consists of finite sequences of approximate moments, one for each unknown measure of (11.45). If one is
interested in statistical properties of the mv solution such as its mean or
its variance, the moments provide the perfect information, at least for sufficiently large d. However, if one is rather interested in properties of the
graph of the entropy solution, a post processing step is required.

An inverse problem. Recovering the graph of the solution
$\{(t, x, y(t, x)) : t \in \mathbf{T}, x \in \mathbf{X}\} \subset \mathbf{T} \times \mathbf{X} \times \mathbf{Y}$ from the moments of the
measure $\nu = \lambda_{\mathbf{T}} \lambda_{\mathbf{X}} \delta_{y(t,x)}$ is an inverse problem whose detailed study is out
of the scope of this chapter; see e.g. Claeys and Sepulchre (2014) in the
context of controlled ODEs. However, we briefly outline here one possible
strategy with a formal justification. It turns out that it works surprising
well in all our examples of the Burgers equation with or without shock.

Let $\mathbf{x} = (t, x, y)$ and $\mathbf{z}_\alpha = \int_{\mathbf{T} \times \mathbf{X} \times \mathbf{Y}} \mathbf{x}^\alpha \hat{d}\nu$ denote the vector of moments
of ν. For any polynomial p in the variables $\mathbf{x}$ with vector of coefficients $\mathbf{p}$
in the monomial basis, it holds

$$\mathbf{p}^\top \mathbf{M}_d(\mathbf{z}) \mathbf{p} = \int p^2 \, d\nu.$$

Consequently, if $\mathbf{p}$ is in the kernel of $M_d(\mathbf{z})$, we have that

$$\int_{\mathbf{T} \times \mathbf{X} \times \mathbf{Y}} p^2 \, d\nu = 0.$$

In other words, the support of the measure is contained in the zero level
set of every polynomial (whose vector of coefficients is) in the kernel of the
moment matrix.

We next show how to approximate this kernel numerically. Since the
moment matrix is positive semidefinite, it has a spectral decomposition

$$\mathbf{M}_d(\mathbf{z}) = \mathbf{P} \mathbf{E} \mathbf{P}^\top \tag{11.46}$$

where $\mathbf{P}$ is an orthonormal matrix whose columns are denoted $\mathbf{p}_i$, $i =
1, 2, \ldots$ and satisfy $\mathbf{p}_i^\top \mathbf{p}_i = 1$ and $\mathbf{p}_i^\top \mathbf{p}_j = 0$ if $i \neq j$, and $\mathbf{E}$ is a diagonal

matrix whose diagonal entries are eigenvalues $e_{i+1} \geq e_i \geq 0$ of the moment matrix. Each column $\mathbf{p}_i$ is the vector of coefficients in the monomial basis of a polynomial p_i in the variables $\mathbf{x}$, so that

$$\mathbf{p}_i^\top \mathbf{M}_d(\mathbf{z}) \mathbf{p}_i = \int p_i^2 \, d\nu = e_i.$$

The following result shows that the measure is concentrated on a sub-level set of an SOS polynomial constructed from the spectral decomposition of the moment matrix.

Lemma 11.8 (Concentration inequality). *Let $r \in \mathbb{N}$ and $\beta > 0$. Define $\gamma = \frac{\sum_{i=1}^r e_i}{\beta}$ and*

$$p_{\mathrm{sos}} = \sum_{i=1}^r p_i^2. \tag{11.47}$$

Then $\nu(\{\mathbf{x} : p_{\mathrm{sos}}(\mathbf{x}) \leq \gamma\}) \geq 1 - \beta$.

The proof of Lemma 11.8 follows readily from the inequality

$$\nu(\{\mathbf{x} : p_{\mathrm{sos}}(\mathbf{x}) > \gamma\}) \leq \frac{\int p_{\mathrm{sos}} \, d\nu}{\gamma} = \beta$$

which holds since ν is a probability measure and p_{sos} is non-negative. Lemma 11.8 justifies the following algorithm, which extracts from a grid the values at which the polynomial p_{sos} is small:

Algorithm 11.1 (Recovery of a solution from moments).
Input: moment matrix $\mathbf{M}_d(\mathbf{z})$ of measure $\nu = \lambda_\mathbf{T} \lambda_\mathbf{X} \delta_{y(t,x)}$, small real $\epsilon > 0$, grid points $(t_i, x_j, y_k)_{i,j,k=1,\ldots,N} \subset \mathbf{T} \times \mathbf{X} \times \mathbf{Y}$.

Step 1 Compute spectral decomposition (11.46) of $\mathbf{M}_d(\mathbf{z})$ and construct SOS polynomial p_{sos} in (11.47) with the largest number of terms r such that $\sum_{i=1}^r e_i < \epsilon$;

Step 2 For each $i, j, k = 1, \ldots, N$, evaluate $p_{i,j,k} := p_{\mathrm{sos}}(t_i, x_j, y_k)$;

Step 3 For each $i, j = 1, \ldots, N$, let $y_{i,j} := y_{k^*}$ where $k^* := \arg\min_k p_{i,j,k}$;

Output: Approximation $(y_{i,j})_{i,j=1,\ldots,N} \subset \mathbf{Y}$ of $y(t,x)$ at grid points $(t_i, x_j)_{i,j=1,\ldots,N} \subset \mathbf{T} \times \mathbf{X}$.

The computational burden of Algorithm 11.1 is modest: an eigenstructure decomposition at Step 1, and grid point evaluations of polynomial p_{sos} at Step 2.

11.4   The Riemann problem for the Burgers equation

For a numerical illustration, we consider the classical Riemann problem (see e.g., Evans (2010)) for a Burgers equation. In particular, we choose the flux

$$f(y) = \frac{1}{4}y^2.$$

The Riemann problem to this conservation law is a Cauchy problem with the following initial condition, piecewise constant with one point of discontinuity:

$$y_0(x) = \begin{cases} l \text{ if } x < 0, \\ r \text{ if } x > 0, \end{cases}$$

where $l, r \in \mathbb{R}$. The solution to the Riemann problem depends strongly on the values of l and r. In particular:

(1) If $l > r$, the shock at the initial condition spreads along the characteristics.
(2) If $l < r$, the solution is not necessarily unique. The entropy condition allows to select the right solution, which is known as a rarefaction wave.

Both cases are interesting from a numerical point of view for their own reasons. In general, the first case is difficult to tackle because of the discontinuity. In general, numerical schemes based on discretization tend to smoothen out the shock. Indeed, recovering numerically the exact point of discontinuity is a challenge for these schemes.

In the second case the solution is continuous, but not necessarily unique. For the Burgers equation, it has been shown that one single entropy condition is sufficient to guarantee uniqueness of the solution DeLellis et al. (2004). To the best of our knowledge, there is no similar result for the uniqueness of entropy mv solutions for Burgers equation with concentrated initial data, except for classical solutions Demoulini et al. (2012).

We present numerical results for both cases. We are going to consider $l, r \in \{0, 1\}$. Following the discussion yielding (11.11), we can assume that the solution takes values only in $\mathbf{Y} = [0, 1]$. The time-space-window on which we consider the solution is $\mathbf{T} = [0, 1]$ and $\mathbf{X} = [L, R] = [-\frac{1}{2}, \frac{1}{2}]$.

Further note that, from the initial condition, we can derive that

$$y(t, L) = l, \quad \forall t \in \mathbf{T}.$$

Moreover, due to positivity of y, the solution on $\mathbf{T} \times \mathbf{X}$ does not depend on the initial condition for $x > \frac{1}{2}$.

Remark on the significance of the numerical results upfront. We need to emphasize that these experiments are by no means conclusive. Our implementation is based on the Matlab interface Gloptipoly3 Henrion et al. (2009) and the SDP solver of MOSEK Dahl (2012). The purpose of the numerical examples is to show that our framework actually works in practice and with a proper implementation might actually provide an alternative to schemes based on discretization.

11.4.1 *Shock waves*

Let $l = 1$ and $r = 0$. As it has been noticed before, with such an initial condition the solution is discontinuous, for all $t > 0$. The unique analytical solution corresponding to this initial condition is

$$y^*(t, x) = \begin{cases} 1 & x > \frac{t}{4}, \\ 0 & x < \frac{t}{4}. \end{cases} \tag{11.48}$$

As an objective function, we choose the default implemented in GloptiPoly, which minimizes the trace of the moment matrix. Since the trace is the convex envelope of the rank on the set of matrices with norm less than one, this is likely to cause early convergence of the moment-SOS hierarchy: low rank solutions correspond to measures supported on sets of zero Lebesgue measure. As in this case their marginal with respect to y is supported on $\{0, 1\}$ we expect this criterion to be appropriate to accelerate convergence. Indeed, for $d = 6$ (i.e. moments of degree up to 12) we end up with the following moments for y:

$$(\mathbf{z}_{0,0,k})_{k=0,1,\ldots} = (1.0000, 0.6250, 0.6250, 0.6250, 0.6250, \ldots)$$

which correspond (up to numerical accuracy) exactly with the moments of the analytic solution.

Localizing the shock. In order to approximate the solution from our approximated moments we follow the path lined out in Section 11.3.3. Applying our algorithm with $\epsilon = 10^{-6}$ yields a polynomial p_{sos} with $r = 54$ terms in the approximate kernel of the moment matrix of size 84, and to the approximated solution represented on Figure 11.1.

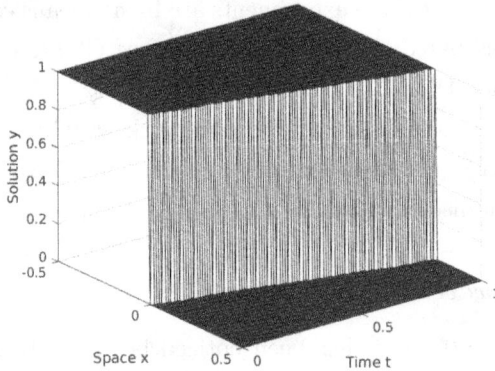

Fig. 11.1   Approximation of the solution $y(t, x)$ obtained with the GMP approach, in the case of a shock.

As already mentioned the computed moments can be used in order to approximate the location of the shock at some given time t. Here we will take $t = 0.75$, consequently the shock is located at exactly $x = 0.1875$. We used a standard Godunov scheme (we refer to Leveque (1992) for more details) to compute the solution up to this time. For space discretization, we took a mesh size of 0.0005 and a consistent discretization in time such that the scheme stays stable. In Table 11.1, we display the obtained values from this approach on an interval around the shock. We can see the typical

Table 11.1 Approximation of $y(0.75, x)$ with Godunov and GMP.

x	0.1850	0.1855	0.1860	0.1865	0.1870	0.1875	0.1880	0.1885
Godunov	0.9999	0.9991	0.9936	0.9580	0.7647	0.2724	0.0123	0.0000
GMP	1.0000	1.0000	1.0000	1.0000	1.0000	0.0000	0.0000	0.0000

behaviour of shock smoothing. In contrast, the values obtained by our GMP approach exactly represent the position of the shock.

11.4.2 Rarefaction waves

Now let $l = 0$ and $r = 1$. As it has been noticed before, with such an initial condition, entropy conditions are crucial to select the right solution, i.e., the solution with a good physical meaning. The analytical entropy solution corresponding to this example is

$$y(t, x) = \begin{cases} 0 & x \leq 0, \\ \frac{2x}{t} & 0 \leq x \leq \frac{t}{2}, \\ 1 & x \geq \frac{t}{2}. \end{cases} \tag{11.49}$$

Numerically implementing all entropy pairs of Kruzkhov is possible (as seen in Section 11.3.1), but heavy. It is known that the entropy $\eta(y) = y^2$ provides all necessary information to make the entropy solution unique for Burgers equation DeLellis et al. (2004). Then, instead of using all Kruzkhov pairs, we propose the following family of entropies in this example:

$$\eta_k(y) = y^k, \qquad \forall k \in \mathbb{N} \tag{11.50}$$

and the corresponding polynomial functions q_k. Note that η is strictly convex on $\mathbf{Y} = [0, 1]$. In particular, we do not have to split the measures in (11.16) into two measures, since there is no absolute value appearing in (11.50). It is neither necessary to introduce a lifting variable as was discussed in Section 11.3.1. Finally, we define the sum over all entropy constraints as an objective function to be maximized.

Solving the relaxation of order $d = 6$ (i.e. moments of degree up to 12), we obtain the following moments for the marginal on y:

$(\mathbf{z}_{0,0,k})_{k=0,1,\dots}$

$= (1.0000, 0.3750, 0.3333, 0.3125, 0.3000, 0.2917, 0.2857, 0.2812, \dots)$

which, again up to numerical accuracy, coincide with the moments of the actual analytic entropy solution. Applying the algorithm from Section 11.3.3 with $\epsilon = 10^{-6}$ yields a polynomial p_{sos} with $r = 48$ terms in the approximate kernel of the moment matrix of size 84, and the approximated solution represented on Figure 11.2.

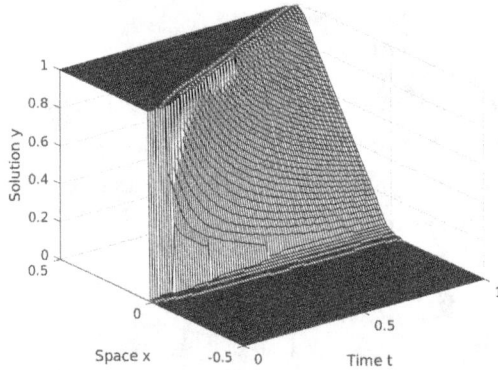

Fig. 11.2 Approximation of the solution $y(t, x)$ obtained with our GMP approach, in the case of a rarefaction wave.

11.5 Notes and sources

Most of the chapter is from Marx et al. (2019). To the best of our knowledge it seems to be the first work where nonlinear PDEs are addressed *without time-space domain discretization* and using convex optimization *with a proof of convergence*. An original early attempt to compute mv solutions of non-linear wave equations with linear programming was reported in Rubio (1997), also in the presence of controls. In Lasserre et al. (2008), the authors have applied the moment-SOS hierarchy to solve optimal control problems of ordinary differential equations. Then it was shown in Henrion and Pauwels (2017) and in Chapter 9 that it provides a sequence of subsolutions converging in norm to the viscosity solution of the Hamilton-Jacobi-Bellmann PDE, a particular non-linear hyperbolic equation. In Mevissen et al. (2011), non-linear PDEs are discretized into large-scale sparse polynomial optimization problems, in turn solved with the moment-SOS hierarchy. More recently, bounds on functionals of solutions were obtained with SOS polynomials for non-linear PDEs arising in fluid dynamics in Chernyshenko et al. (2014) and in Goluskin and Fantuzzi (2018) for the non-linear Kuramoto-Sivashinsky PDE. However these works focus only on the dual SOS problem, and they provide bounds with no convergence guarantee. In particular, they do not exploit the primal formulation of the problem on moments, which we believe to be crucial for convergence analysis. In their recent work Fjordholm et al. (2016) compute mv solutions for the equations of compressible and incompressible inviscid fluid dynamics, with the help of discretization algorithms based on Monte Carlo methods. Even more recently Brenier (2017) has proposed a convex formulation for the classical solution to nonlinear hyperbolic PDEs and he proves that the entropy solution to the Burgers equation might be recovered also via this optimization problem. However this paper does not provide a numerical scheme. In a concurrent work, Korda et al. (2018) propose to use the Moment-SOS hierarchy in a much more general setting of a controlled polynomial PDE. However at such a level of generality, there is no proof that the numerical scheme will converge to an appropriate solution of the PDE. For more references on previous attempts to use convex optimization for solving and controlling PDEs, the reader is referred to the introduction of Korda et al. (2018).

Finally one may wonder whether it is possible to extend the Moment-SOS hierarchy to other nonlinear PDEs and also to some inverse problems for PDEs.

- The class of nonlinear parabolic PDEs. One interesting feature of these equations is that they *regularize* the solution, whatever is the initial condition. Therefore, as it is done for ODEs in Lasserre et al. (2008), it might be possible to define test functions depending on the solution to the parabolic equation and then define an occupation measure associated to the latter. This together with the relaxed control theory surveyed in Fattorini (1999) might be instrumental to solve optimal control problems for nonlinear parabolic equations.

- The Burgers equation is irreversible. Roughly speaking, given a terminal condition $y(T, x)$ with $T > 0$, there exists a continuum of initial conditions yielding $y(T, x)$, see e.g., Gosse and Zuazua (2017). Such a continuum can be described with measures and, hence, our linear formulation might be useful to solve such inverse problems, extending to PDEs what was developed in Henrion and Korda (2014) for ODEs.

Bibliography

Brenier, Y. (2017). *Solution by convex minimization of the Cauchy problem for hyperbolic systems of conservation laws with convex entropy.* arXiv:1710.03754.

Brezis, H. (2010). *Functional analysis, Sobolev spaces and partial differential equations.* Springer.

Chernyshenko, S. I., Goulart, P., Huang, D. and Papachristodoulou, A. (2014) *Polynomial sum of squares in fluid dynamics: a review with a look ahead.* Phil. Trans. R. Soc. A, 372(2020):20130350.

Claeys, M. and Sepulchre, R. (2014). *Reconstructing trajectories from the moments of occupation measures.* Proc. IEEE Conf. on Decision and Control.

Dahl, J. (2012). *Extending the conic optimizer in MOSEK with semidefinite cones.* Proc. Intl. Symp. Math. Prog., Berlin.

Dafermos, C. M. (2006). *Hyperbolic conservation laws in continuum physics.* Springer.

DeLellis, C., Otto, F. and Westdickenberg, M. (2004). *Minimal entropy conditions for Burgers equation.* Quarterly of applied mathematics 62(4):687–700.

Demoulini, S., Stuart, D. M. A. and Tzavaras, A. E. (2012). *Weak-strong uniqueness of dissipative measure-valued solutions for polyconvex elastodynamics.* Archive for Rational Mechanics and Analysis 205(3):927–961.

Després, B. and Lagoutière, F.(2001). *Contact discontinuity capturing schemes for linear advection and compressible gas dynamics.* Journal of Scientific Computing 16(4):479–524.

DiPerna, R. J. (1985) *Measure-valued solutions to conservation laws.* Archive for Rational Mechanics and Analysis 88(3):223–270.

Evans, L. C. (2010). *Partial differential equations.* American Mathematical Society.

Fattorini, H. O. (1999). *Infinite dimensional optimization and control theory.* Cambridge University Press.

Feireisl, E., Lukácová-Medvid'ová, M. and Mizerová, H. (2019). *Convergence of finite volume schemes for the Euler equations via dissipative measure-valued solutions.* arXiv:1803.08401, to appear in Foundations of Computational Mathematics.

Fjordholm, U. S., Mishra, S. and Tadmor, E. (2016). *On the computation of measured-valued solutions.* Acta Numerica 25:567–679.

Fjordholm, U. S., Käppeli, R., Mishra, S., and Tadmor, E. (2017). *Construction of approximate entropy measure-valued solutions for hyperbolic systems of conservation laws.* Foundations of Computational Mathematics 17(3):763–827.

Godunov, S. K. (1959). *A difference method for numerical calculation of discontinuous solutions of the equations of hydrodynamics.* Matematicheskii Sbornik 89(3):271–306.

Goluskin, D. and Fantuzzi, G.(2018). Bounds on mean energy in the Kuramoto-Sivashinsky equation computed using semidefinite programming. arXiv:1802.08240.

Gosse, L. and Zuazua, E. (2017). *Filtered gradient algorithms for inverse design problems of one-dimensional Burgers equation.* Pages 197–227 in Gosse, L. and Natalini, G. (Editors). Innovative algorithms and analysis. SINDAM Series, Springer.

Handelman, D. (1988). *Representing polynomials by positive linear functions on compact convex polyhedra.* Pacific J. Math. 132(1):35–62.

Henrion, D. and Korda, M. (2014). *Convex computation of the region of attraction of polynomial control systems.* IEEE Trans. Autom. Control 59(2):297–312.

Henrion, D., Lasserre, J. B., and Löfberg, J. (2009). *Gloptipoly 3: moments, optimization and semidefinite programming.* Optimization Methods & Software 24(4-5):761–779.

Henrion, D. and Pauwels, E. (2017). *Linear conic optimization for nonlinear optimal control.* Pages 121–134 in Ahmed, S., Anjos, M. and Terlaky, T. (Editors). Advances and Trends in Optimization with Engineering Applications. SIAM.

Korda, M., Henrion, D. and Lasserre, J. B. (2018). *Moments and convex optimization for analysis and control of nonlinear partial differential equations.* arXiv:1804.07565.

Kruzkhov, S. N. (1970). *First order quasilinear equations in several independent variables.* Mathematics of the USSR-Sbornik 10(2):217.

Lasserre, J. B. (2010). *Moments, positive polynomials and their applications.* Imperial College Press.

Lasserre, J. B., Henrion, D., Prieur, C. and Trélat, E. (2008). *Nonlinear optimal control via occupation measures and LMI relaxations.* SIAM Journal on Control and Optimization 47(4):1643–1666.

Lax, P. (1971). *Shock waves and entropy.* Pages 603–634 in Zarantonello, E. H. (Editor). Contributions to nonlinear functional analysis. Elsevier.

LeVeque, R. J. (1992). *Numerical methods for conservation laws.* Lectures in Mathematics, ETH Zürich.

Marx, S., Weisser, T., Henrion, D. and Lasserre, J. B. (2019). *A moment approach for entropy solutions to nonlinear hyperbolic PDEs.* Math. Control and Related Fields 10(1):113–140.

Mevissen, M., Lasserre, J. B. and Henrion, D. (2011). *Moment and SDP relaxation techniques for smooth approximations of problems involving nonlinear differential equations.* Proc. IFAC World Congress on Automatic Control.

Málek, J., Necas, J., Rokyta, M., and Ruzicka, M. (1996). *Weak and measure-valued solutions to evolutionary PDEs.* CRC Press.

Panov, E. Y. (1994). *Uniqueness of the solution of the Cauchy problem for a first order quasilinear equation with one admissible strictly convex entropy.* Mathematical Notes 55(5):517–525.

Putinar, M. (1993). *Positive polynomials on compact semi-algebraic sets.* Indiana University Mathematics Journal 42(3):969–984.

Rubio, J. (1997). *The global control of shock waves.* Pages 355–369 in
Grosser, M., Hörmann, G., Kunzinger, M., and Oberguggenberger, L.
(Editors). Nonlinear theory of generalized functions. Erwin Schrödinger
Institute, Vienna.

Whitham, G. B. (2011). *Linear and nonlinear waves.* John Wiley & Sons.

Chapter 12

Miscellaneous

We provide some additional material: I. How to use a certain structured sparsity pattern in large scale problems to provide a sparsity-adapted version of the Moment-SOS hierarchy. II. Introduce the Christoffel function, a tool of approximation theory with striking properties that may be useful in other contexts.

12.1 Scalability and sparsity

Scalability of the Moment-SOS hierarchy is an important issue. Indeed the Moment-SOS approach as described in previous chapters does not scale well with the problem size and therefore the application of its standard version is limited to problems of small to medium size only.

Fortunately large scale problems often exhibit special structures that can be exploited for efficient computational purposes. Sparsity and symmetries are two such features. In this section we briefly describe how to take advantage of some structured sparsity when it is present in the initial problem description.

12.1.1 *Structured sparsity pattern*

Let $\Omega \subset \mathbb{R}^n$ be a compact basic semi-algebraic set defined by:

$$\Omega := \{\mathbf{x} \in \mathbb{R}^n : g_j(\mathbf{x}) \geq 0, \quad j = 1, \ldots, m\} \tag{12.1}$$

for some polynomial $(g_j) \subset \mathbb{R}[\mathbf{x}]$.

The sparsity pattern

We assume that the data of problem **P** satisfy the following structured sparsity pattern:

- The index set $\mathbf{I}_0 := \{1, 2, \ldots, n\}$ where n is potentially "large", is a union $\mathbf{I}_0 = \bigcup_{k=1}^{p} \mathbf{I}_k$ (with possible overlaps) where $\#\mathbf{I}_k \ll n$ for all k.
- With Ω as in (12.1), for every $j = 1, \ldots, m$, $g_j \in \mathbb{R}[x_i : i \in \mathbf{I}_{k(j)}]$ for some $1 \leq k(j) \leq p$.
- Ω is compact and we may and will assume that for each $k = 1, \ldots, p$, there exists $j_k \in \{1, \ldots, m\}$ such that $g_{j_k} = M - \sum_{i \in \mathbf{I}_k}^{n} x_i^2$ (for some $M > 0$ sufficiently large). (One may always include such a redundant constraint in the description of Ω.)

We also consider the following assumption on the subsets $(\mathbf{I}_k)$.

Assumption 12.1. The subsets $(\mathbf{I}_k)$ satisfy the so-called *Running Intersection Property*:

$$\forall k = 2, \ldots, p: \quad \mathbf{I}_k \cap \left(\bigcup_{j=1}^{k-1} \mathbf{I}_j \right) \subseteq \mathbf{I}_s \tag{12.2}$$

for some $s \leq k - 1$.

A sparse version of Putinar's Positivstellensatz. The following result provides a sparsity-adapted version of Putinar's Theorem 1.9.

Theorem 12.1. *(Lasserre (2006)) Let Assumption 12.1 hold and let $f \in \mathbb{R}[\mathbf{x}]$ be in the form $f = \sum_{k=1}^{p} f_k$ with $f_k \in \mathbb{R}[x_i : i \in \mathbf{I}_k]$, $k = 1, \ldots, p$. If f is (strictly) positive on Ω then*

$$f = \sum_{k=1}^{p} \left(\sigma_{k0} + \sum_{j:g_j \in \mathbb{R}[x_i:i\in\mathbf{I}_k]} \sigma_{kj}\, g_j \right), \tag{12.3}$$

for some SOS polynomials $\sigma_{kj} \in \Sigma[x_i : i \in \mathbf{I}_k]$.

Notice that in (12.3) the SOS weights σ_{kj} are polynomials in the variables x_i, $i \in \mathbf{I}_k$, only. This is in contrast to Theorem 1.9 where the SOS weights are polynomials in *all* the variables x_i, $i \in \mathbf{I}_0$.

Therefore in all the applications considered in previous chapters, Theorem 12.1 can be used when positivity certificates are needed, as soon as a sparsity pattern (as above) is identified in the problem description.

12.1.2 A sparse Moment-SOS hierarchy for optimization

Consider the polynomial optimization problem

$$\mathbf{P}: \quad f^* = \min_{\mathbf{x}}\{\, f(\mathbf{x}) : \mathbf{x} \in \boldsymbol{\Omega} \},$$

with $\boldsymbol{\Omega}$ as in (12.1), already encountered in Chapter 2. Define $d_j := \lceil \deg(g_j)/2 \rceil$, for $j = 1,\ldots,m$ (and $d_0 := 0$), and for every $d \geq \max_j d_j$, consider the semidefinite program:

$$
\begin{aligned}
\rho_d^{\text{sparse}} = \max_{\sigma_{kj},\lambda} \Big\{ \lambda : f - \lambda = \sum_{k=1}^{p}\Big(&\sigma_{k0} + \sum_{j:g_j \in \mathbb{R}[x_i : i \in \mathbf{I}_k]} \sigma_{kj}\, g_j \Big) \\
&\sigma_{kj} \in \Sigma[x_i : i \in \mathbf{I}_k]_{d-d_j}, \quad j = 0,\ldots,m \Big\}
\end{aligned}
\tag{12.4}
$$

The sequence $(\rho_d^{\text{sparse}})_{d\in\mathbb{N}}$ is monotone non decreasing and the SOS-hierarchy (12.4) is the sparse version of the (dense) version:

$$
\begin{aligned}
\rho_d = \max_{\sigma_j,\lambda} \Big\{ \lambda : f - \lambda &= \sigma_0 + \sum_{j=1}^{m} \sigma_j\, g_j\,; \\
\sigma_j &\in \Sigma[\mathbf{x}]_{d-d_j}\ j = 0,\ldots,m \Big\}
\end{aligned}
\tag{12.5}
$$

Corollary 12.2. *With $\boldsymbol{\Omega}$ as in (12.1), let Assumption 12.1 hold and let $f = \sum_{k=1}^{p} f_k$ with $f_k \in \mathbb{R}[x_i : i \in \mathbf{I}_k]$, $k = 1,\ldots,p$. Then the sequence $(\rho_d)_{d\in\mathbb{N}}$ is monotone non-decreasing and $\rho_d^{\text{sparse}} \to f^*$ as $d\to\infty$.*

To compare the computational cost involved in the sparse version (12.4) versus the dense version (12.5), let $\tau := \max_k \#\mathbf{I}_k$.

• In (12.5) one has $m + 1$ SOS polynomials in n variables of degree at most $2d$, i.e., involving $(m+1)\binom{n+2d}{n}$ coefficients, and $m + 1$ semidefinite constraints of size $\binom{n+d}{n}$.

• In (12.4) one has $m + 1$ SOS polynomials in at most τ variables of degree at most $2d$, i.e., involving $(m+1)\binom{\tau+2d}{\tau}$ coefficients, and $m + 1$ semidefinite constraints of size $\binom{\tau+d}{\tau}$.

So for fixed n and when d varies, step-d in the dense SOS-hierarchy involves $(m+1)O((2d)^n)$ variables (resp. $(m+1)$ semidefinite constraints of size $O(d^n)$), to compare with $(m+1)O((2d)^\tau)$ (resp. $(m+1)O(d^\tau)$) in the sparse version.

Therefore when τ is relatively small (e.g. say $\tau \leq 6,7$) then one may handle polynomial optimization problems **P** with quite a large number of variables (e.g. $n > 2000$) for which sometimes even the first step in the SOS-hierarchy ($d = 1$) cannot be implemented.

12.1.3 *An alternative sparse Positivstellensatz*

For every $k = 1, \ldots, p$, let $J_k = \{j \in \{1, \ldots, m\} : g_j \in \mathbb{R}[x_i : i \in \mathbf{I}_k]\}$ and $a_k := \# J_k$.

Assumption 12.2. Let Ω be as in (12.1). (i) For each $k = 1, \ldots, p$, some polynomials $g_{j_1}, \ldots, g_{j_t} \in \mathbb{R}[x_i : i \in \mathbf{I}_k]$ generate $\mathbb{R}[x_i : i \in \mathbf{I}_k]$. (If not it suffices to add some redundant constraints of the form $x_i \leq M_k$, $i \in \mathbf{I}_k$, where M_k is sufficiently large.)

(ii) Possibly after scaling, $0 \leq g_j \leq 1$ on Ω, for every $j = 1, \ldots, m$.

Theorem 12.3. *(Weisser et al. (2018)) Let Assumption 12.1 and 12.2 hold and let $1 \leq s \in \mathbb{N}$ be fixed. Let $f = \sum_{k=1}^{p} f_k$ with $f_k \in \mathbb{R}[x_i : i \in \mathbf{I}_k]$, $k = 1, \ldots, p$. If f is (strictly) positive on Ω then*

$$f = \sum_{k=1}^{p} \left(\sigma_k + \sum_{\alpha,\beta \in \mathbb{N}^{a_k}} c_{k,\alpha\beta} \prod_{j \in J_k} g_j^{\alpha_j} (1 - g_j)^{\beta_j} \right) \qquad (12.6)$$

for some finitely many positive coefficients $c_{k,\alpha\beta}$ and SOS polynomials $\sigma_k \in \Sigma[x_i : i \in \mathbf{I}_k]$ of degree at most $2s$.

In some cases this positivity certificate might be more interesting than (12.3) because in the former the degree of the SOS polynomials involved is fixed in advance ($= 2s$). Therefore instead of (12.4) one may consider the alternate Moment-SOS hierarchy:

$$\rho_d^{\text{sparse}} = \max_{\sigma_k, c_{k,\alpha\beta}, \lambda} \left\{ \lambda : f - \lambda = \sum_{k=1}^{p} (\sigma_k \right.$$
$$+ \sum_{\alpha,\beta \in \mathbb{N}^{a_k}} c_{k,\alpha\beta} \prod_{j \in J_k} g_j^{\alpha_j} (1 - g_j)^{\beta_j}) \qquad (12.7)$$
$$\left. c_{k,\alpha\beta} \geq 0; \ \sigma_k \in \Sigma[x_i : i \in I_k]_s \right\}.$$

Notice that in (12.7) (and in contrast to (12.4)), the size of the $m+1$ semidefinite constraints (each associated with the SOS polynomials σ_k) is fixed ($\leq \binom{\tau+s}{\tau}$) and does not depend on d. On the other hand, the number of variables $c_{k,\alpha\beta}$ can be quite large ($> O((2\tau)^d)$) and so this strategy can be interesting if τ is small (e.g. $\leq 3, 4$). For more details on properties of the SOS-hierarchy (12.7) the interested reader is referred to Weisser et al. (2018).

12.2 A distinguished SOS: The Christoffel polynomial

In this section we mention some properties of a distinguished SOS polynomial, already encountered in Chapter 6. We start with a compact domain $\Omega \subset \mathbb{R}^n$ and a finite Borel measure μ on Ω such that its moment matrix is positive definite ($\mathbf{M}_d(\mu) \succ 0$) for all $d = 0, 1, \dots$. Recall that $\mathbf{v}_d(x) = (x^\alpha)_{\alpha \in \mathbb{N}_d^n}$ is the vector of monomials up to order d (assuming a certain ordering of monomials).

Recall that the support of μ (denoted supp(μ)) is the largest closed set whose measure of its complement vanishes. So one may and will assume that $\Omega = \text{supp}(\mu)$.

Definition 12.1. Let $c_d \in \Sigma[\mathbf{x}]_d$ be the SOS defined by:

$$\mathbf{x} \mapsto c_d(\mathbf{x}) := \mathbf{v}_d(\mathbf{x})^T \mathbf{M}_d(\mu)^{-1} \mathbf{v}_d(\mathbf{x}), \quad d \in \mathbb{N}. \qquad (12.8)$$

Its reciprocal c_d^{-1} is called the Christoffel function associated with μ. Whence we call c_d (of degree $2d$) the Christoffel polynomial associated with μ.

It turns out that:

$$c_d(\mathbf{x}) = \sum_{\alpha \in \mathbb{N}_d^n} P_\alpha(\mathbf{x})^2, \qquad\qquad (12.9)$$

where the $(P_\alpha)_{\alpha \in \mathbb{N}^n}$ form a family of polynomials that are orthonormal with respect to μ, that is, the P_α's satisfy:

$$\int_\Omega P_\alpha P_\beta \, d\mu = \delta_{\alpha=\beta}, \quad \alpha, \beta \in \mathbb{N}^n. \qquad\qquad (12.10)$$

Let $(\mathbf{x}, \mathbf{y}) \mapsto K_d(\mathbf{x}, \mathbf{y})$ be the *kernel* defined by:

$$(\mathbf{x}, \mathbf{y}) \mapsto K_d(\mathbf{x}, \mathbf{y}) := \sum_{\alpha \in \mathbb{N}_d^n} P_\alpha(\mathbf{x}) \, P_\alpha(\mathbf{y}), \qquad \mathbf{x}, \mathbf{y} \in \mathbb{R}^n,$$

with the P_α's as in (12.10). Then K_d is a *reproducing kernel* for the Hilbert subspace $\mathbb{R}[\mathbf{x}]_d \subset L_2(\mu)$ because

$$P_\alpha(\mathbf{x}) = \int_\Omega K_d(\mathbf{x}, \mathbf{y}) \, P_\alpha(\mathbf{y}) \, d\mu(\mathbf{y}), \quad \alpha \in \mathbb{N}_d^n$$

and so by linearity:

$$f(\mathbf{x}) = \int_\Omega K_d(\mathbf{x}, \mathbf{y}) \, f(\mathbf{y}) \, d\mu(\mathbf{y}), \quad \forall f \in \mathbb{R}[\mathbf{x}]_d.$$

The following states an important extremal property (or alternative definition) of c_d:

Proposition 12.4. *Let $c_d \in \Sigma[\mathbf{x}]$ be as in Definition 12.1. Then:*

$$\frac{1}{c_d(\boldsymbol{\xi})} = \min_{p \in \mathbb{R}[\mathbf{x}]_d} \left\{ \int_\Omega p^2 \, d\mu : p(\boldsymbol{\xi}) = 1 \right\}, \quad \forall \boldsymbol{\xi} \in \mathbb{R}^n. \qquad (12.11)$$

The optimal solution $p^* \in \mathbb{R}[\mathbf{x}]_d$ in (12.11) is just the polynomial

$$\mathbf{x} \mapsto p_d^*(\mathbf{x}) := \frac{1}{c_d(\boldsymbol{\xi})} K_d(\mathbf{x}, \boldsymbol{\xi}) = \frac{K_d(\mathbf{x}, \boldsymbol{\xi})}{K_d(\boldsymbol{\xi}, \boldsymbol{\xi})}, \quad \mathbf{x} \in \mathbb{R}^n.$$

Lemma 12.5. *For every $\boldsymbol{\xi} \notin \Omega$, $\lim_{d \to \infty} \binom{n+d}{n} c_d(\boldsymbol{\xi})^{-1} = 0$.*

So there is a simple way to "get an idea" of the support of μ from the sole knowledge of it moments up to order $2d$.

- Form the moment matrix $\mathbf{M}_d(\mu)$
- Compute the SOS polynomial $c_d \in \Sigma[\mathbf{x}]_d$

Points $\boldsymbol{\xi} \in \mathbb{R}^n$ outside the sub-level set $\{\mathbf{x} : \frac{c_d(\mathbf{x})}{\binom{n+d}{n}} \leq \rho\}$ with $\rho > 1$, are likely to be outside the support of μ.

Illustrative example. We next show how in 2D-examples the geometric shape of a cloud of points can be captured by simple computation of the empirical moment matrix. So let μ be the empirical probability measure associated with a cloud of N points in $\mathbb{R}^2$. In Figure 12.1 such a cloud of $N = 1000$ points is plotted, and most points of the cloud are contained in sublevel set $\{\mathbf{x} : c_d(\mathbf{x}) \leq \binom{n+d}{n}\}$ (with $d = 4$) where the level set $\{\mathbf{x} : c_d(\mathbf{x}) = \binom{n+d}{n}\}$ is plotted in red. One may observe that the sublevel set $\{\mathbf{x} : c_d(\mathbf{x}) \leq \binom{n+d}{n}\}$ captures the geometric shape of the cloud quite well even though $d = 4$ is small. One may also check that this is indeed the case for several clouds of points in $\mathbb{R}^2$ with quite different shapes as illustrated in Figure 12.2 where for each example of cloud, the corresponding level sets $\{\mathbf{x} : c_d(\mathbf{x}) = \binom{n+d}{n}\}$ are displayed in red for several (small) values of $d = 3, 4, 5$.

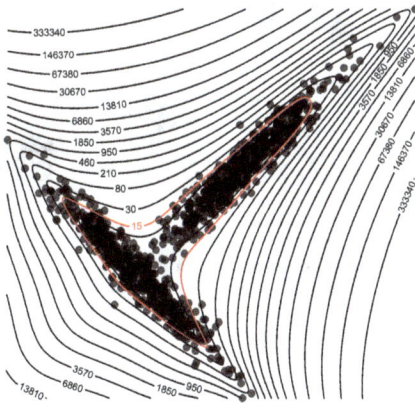

Fig. 12.1   In red the level set $\{\mathbf{x} : c_d(\mathbf{x}) = \binom{n+d}{n}\}$

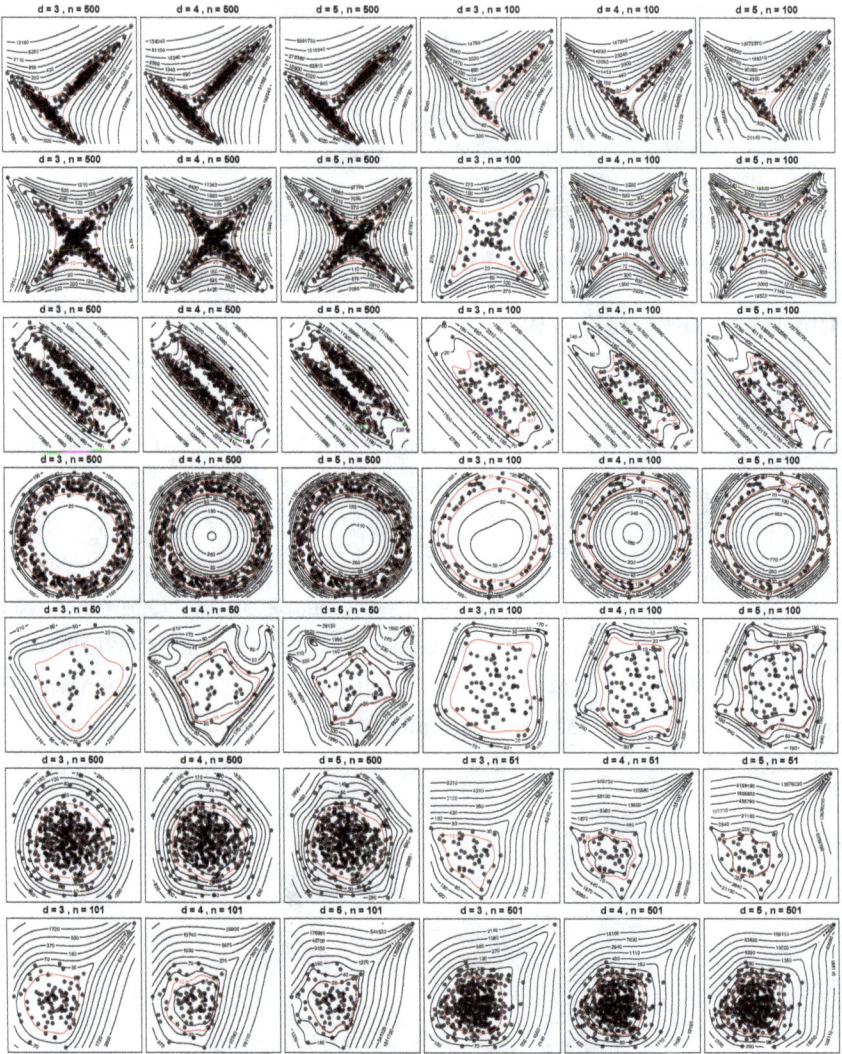

Fig. 12.2 Examples of 2D-clouds: in red the level sets $\{\mathbf{x} : c_d(\mathbf{x}) = \binom{n+d}{n}\}$, $d = 3, 4, 5$

12.3 Notes and sources

The Moment-SOS hierarchy is a very powerful numerical scheme but in its dense form (12.5) it is computationally expensive. In particular, it does not scales well with the problem size. Therefore in view of the present status of semidefinite solvers, its application is limited to problems of modes size. On the other hand it is often the case that large scale problems exhibit some symmetries and/or some structured sparsity pattern which can be exploited. The resulting appropriate semidefinite relaxations that take such features into account (e.g. as in (12.2) of (12.7)) are of much smaller size; see for instance Lasserre (2006), Bugarin et al. (2016), Weisser et al. (2018).

The Christoffel function has a rich history and is essentially concerned with the theory of approximation and orthogonal polynomials; see e.g. Dunkl and Xu (2001). If the Christoffel function is a well-known tool in approximation theory its is much less known in other fields like statistics and optimization. One important (and difficult) issue is to provide conditions for the pointwise convergence of the Christoffel function as the degree d increases. Such conditions have been provided in a number of interesting cases and the limit has even been identified in some particular univariate and multivariate cases; it involves the density of an *equilibrium measure* which is intrinsic to the support of the considered measure. For more details and an historical account on the Christoffel function the interested reader is referred to Bos (1994); Bos et al. (1998); Gustafsson et al. (2009); Króo and Lubinsky (2013); Máté and Nevai (1980); Máté et al. (1991); Nevai (1986); Totik (2000); Xu (1996, 1999).

It turns out that this simple tool can be very useful in some important applications of data analysis and machine learning. The interested reader is referred to Lasserre and Pauwels (2016, 2019), and Pauwels et al. (2019) for several examples of such applications.

Bibliography

Bos, L. (1994). *Asymptotics for the Christoffel function for Jacobi like weights on a ball in* $\mathbb{R}^m$, New Zealand J. Math. **23**, pp. 109–116.

Bos, L., Della Vecchia, B., and Mastroianni, G. (1998). *On the asymptotics of Christoffel functions for centrally symmetric weight functions on the ball in* $\mathbb{R}^d$, Rend. Circ. Mat. Palermo **2**, pp. 277–290.

Bugarin, F., Henrion, D., and Lasserre, J. B. (2016) *Minimizing the sum of many rational functions*, Math. Program. Comput. **8**, pp. 83–111.

Dunkl, C., and Xu, Y. (2001). *Orthogonal polynomials of several variables* (Cambridge University Press, Cambridge).

Gustafsson, B., Putinar, M., Saff, E., and Stylianopoulos, N. (2009). *Bergman polynomials on an archipelago: estimates, zeros and shape reconstruction*, Adv. Math. **222**, pp. 1405–1460.

Króo, A., and Lubinsky, D. S. (2013). *Christoffel functions and universality on the boundary of the ball*, Acta. Math. Hungarica **140**, pp. 117–133.

Lasserre, J. B., and Pauwels, E. (2016). *Sorting out typicality with the inverse moment matrix sos polynomial*, In *Advances in Neural Information Processing Systems 29* (NIPS 2016, Barcelona), D. D. Lee and M. Sugiyama and U. V. Luxburg and I. Guyon and R. Garnett, Eds., Curran Associates, Inc. 2016, pp. 190–198.

Lasserre, J. B., and Pauwels, E. (2019). *The empirical Christoffel function with applications in data analysis*, Adv. Comp. Math. **45**, pp. 1439–1468.

Pauwels, E., Putinar, M., and Lasserre, J. B. (2018). *Data analysis from empirical moments and the Christoffel function*, Found. Comp. Math. to appear. arXiv:1810.08480.

Henrion D., Lasserre, J. B., Lofberg J. (2009). *Gloptipoly 3: moments, optimization and semidefinite programming*, Optim. Methods and Softwares **24**, pp. 761–779.

Lasserre, J. B. (2006). *Convergent SDP-relaxations in polynomial optimization with sparsity*, SIAM J. Optim. **17**, pp. 822–843.

Máté, A., and Nevai, P. (1980). *Bernstein's Inequality in L_p for $0 < p < 1$ and $(C, 1)$ Bounds for Orthogonal Polynomials*, Ann. Math. **111**, pp. 145–154.

Máté, A., Nevai, P., and Totik, V. (1991). *Szego's extremum problem on the unit circle*, Ann. Math., 433–453.

Nevai, P. (1986). *Geza Freud, orthogonal polynomials and Christoffel functions. A case study*, J. Approx. Theory **48**, pp. 3–167.

Totik, V. (2000). *Asymptotics for Christoffel functions for general measures on the real line*, J. Anal. Math. **81**, pp. 283–303.

Weisser, T., Lasserre, J. B., and Toh, K. (2018). *A bounded degree SOS hierarchy for large scale polynomial optimization with sparsity*, Math. Program. Comput. **10**, pp. 1–32.

Xu, Y. (1996). *Asymptotics for orthogonal polynomials and Christoffel functions on a ball*, Methods and Appl. Anal. **3**, pp. 257–272.

Xu, Y. (1999). *Asymptotics of the Christoffel Functions on a Simplex in $\mathbb{R}^d$*, J. Approx. Theory **99**, pp. 122–133.

Glossary

- $\mathbb{N}$, the set of natural numbers.
- $s(d) = \binom{n+d}{n}$
- $\mathbb{Z}$, the set of integers.
- $\mathbb{Q}$, the set rational numbers.
- $\mathbb{R}$, the set of real numbers.
- $\mathbb{R}_+$, the set of nonnegative real numbers.
- $\mathbb{C}$, the set of complex numbers.
- $\leq$, less than or equal to
- $\leqq$, inequality "$\leq$" or equality "$=$"
- $\mathbf{A}$, matrix in $\mathbb{R}^{m \times n}$,
- $\mathbf{A}_j$, column j of matrix $\mathbf{A}$.
- $\mathbf{A} \succeq 0 \, (\succ 0)$, $\mathbf{A}$ is positive semidefinite (definite)
- x, scalar $x \in \mathbb{R}$
- $\mathbf{x}$, vector $\mathbf{x} = (x_1, \ldots, x_n) \in \mathbb{R}^n$
- $\boldsymbol{\alpha}$, vector $\boldsymbol{\alpha} = (\alpha_1, \ldots, \alpha_n) \in \mathbb{N}^n$
- $|\boldsymbol{\alpha}|, = \sum_{i=1}^n \alpha_i$ for $\boldsymbol{\alpha} \in \mathbb{N}^n$.
- $\mathbb{N}_d^n$, $\subset \mathbb{N}^n$, the set $\{\boldsymbol{\alpha} \in \mathbb{N}^n : |\boldsymbol{\alpha}| \leq d\}$
- $\mathbf{x}^{\boldsymbol{\alpha}}$, vector $\mathbf{x}^{\boldsymbol{\alpha}} = (x_1^{\alpha_1} \cdots x_n^{\alpha_n})$, $\mathbf{x} \in \mathbb{C}^n$ or $\mathbf{x} \in \mathbb{R}^n$, $\boldsymbol{\alpha} \in \mathbb{N}^n$.
- $\mathbb{R}[x]$; ring of real univariate polynomials
- $\mathbb{R}[\mathbf{x}], = \mathbb{R}[x_1, \ldots, x_n]$, ring of real multivariate polynomials
- $\mathcal{P}(\mathcal{X})_d$: space of polynomials of degree at most d, nonnegative on $\mathcal{X}$.
- $(\mathbf{x}^{\boldsymbol{\alpha}})$, canonical monomial basis of $\mathbb{R}[\mathbf{x}]$
- $V_{\mathbb{C}}(I) \subset \mathbb{C}^n$, the algebraic variety associated with an ideal $I \subset \mathbb{R}[\mathbf{x}]$
- $\sqrt{I}$, the radical of an ideal $I \subset \mathbb{R}[\mathbf{x}]$
- $\sqrt[\mathbb{R}]{I}$, the real radical of an ideal $I \subset \mathbb{R}[\mathbf{x}]$
- $I(V(I))$, $\subset \mathbb{C}^n$, the vanishing ideal $\{f \in \mathbb{R}[\mathbf{x}] : f(\mathbf{z}) = 0 \ \forall \mathbf{z} \in V_{\mathbb{C}}(I)\}$.
- $V_{\mathbb{R}}(I)$, $\subset \mathbb{R}^n$, the real variety associated with an ideal $I \subset \mathbb{R}[\mathbf{x}]$

- $I(V_{\mathbb{R}}(I))$, $\subset \mathbb{R}[\mathbf{x}]$, the real vanishing ideal $\{f \in \mathbb{R}[\mathbf{x}] : f(\mathbf{x}) = 0 \; \forall \mathbf{x} \in V_{\mathbb{R}}(I)\}$.
- $\mathbb{R}[\mathbf{x}]_t$, $\subset \mathbb{R}[\mathbf{x}]$, real multivariate polynomials of degree at most t
- $(\mathbb{R}[\mathbf{x}])^*$, the vector space of linear forms on $\mathbb{R}[\mathbf{x}]$
- $(\mathbb{R}[\mathbf{x}]_t)^*$, the vector space of linear forms on $\mathbb{R}[\mathbf{x}]_t$
- $\mathbf{y} = (y_\alpha) \subset \mathbb{R}$, moment sequence indexed in the canonical basis of $\mathbb{R}[\mathbf{x}]$
- $\mathbf{M}_d(\mathbf{y})$, moment matrix of order d associated with the sequence $\mathbf{y}$
- $\mathbf{M}_d(g\,\mathbf{y})$, localizing matrix of order d associated with the sequence $\mathbf{y}$ and $g \in \mathbb{R}[\mathbf{x}]$
- $P(g)$, $\subset \mathbb{R}[\mathbf{x}]$, preordering generated by the polynomials $(g_j) \subset \mathbb{R}[\mathbf{x}]$
- $Q(g)$, $\subset \mathbb{R}[\mathbf{x}]$, quadratic module generated by the polynomials $(g_j) \subset \mathbb{R}[\mathbf{x}]$
- $\mathcal{M}(\mathbf{K})_d$, set of finite sequences $\mathbf{y} \in \mathbb{R}^{s(d)}$ with a representing measure on $\mathbf{K}$.
- $\mathrm{co}\,\mathbf{X}$, convex hull of $\mathbf{X} \subset \mathbb{R}^n$
- $\mathscr{B}\mathbf{X})$ (resp. $\mathscr{B}(\mathbf{X})_+$) space of bounded (resp. bounded nonnegative) measurable functions on $\mathbf{X}$.
- $\mathscr{C}(\mathbf{X})$ (resp. $\mathscr{C}(\mathbf{X})_+$), space of bounded (resp. bounded nonnegative) continuous functions on $\mathbf{X}$.
- $\mathscr{M}(\mathbf{X})$, vector space of finite signed Borel measures on $\mathbf{X} \subset \mathbb{R}^n$
- $\mathscr{M}(\mathbf{X})_+$, $\subset \mathscr{M}(\mathbf{X})$, space of finite Borel measures on $\mathbf{X} \subset \mathbb{R}^n$
- $\mathscr{P}(\mathbf{X})$, $\subset \mathscr{M}(\mathbf{X})_+$, space of Borel probability measures on $\mathbf{X} \subset \mathbb{R}^n$
- $L_p(\mathbf{X}, \mu)$, Banach of functions on $\mathbf{X} \subset \mathbb{R}^n$ such that $(\int_{\mathbf{X}} |f|^p d\mu)^{1/p} < \infty$, $1 \le p < \infty$.
- $L_\infty(\mathbf{X}, \mu)$, Banach space of measurable functions on $\mathbf{X} \subset \mathbb{R}^n$ such that $\|f\|_\infty := \mathrm{ess}\,\sup |f| < \infty$.
- $\sigma(\mathcal{X}, \mathcal{Y})$, weak topology on $\mathcal{X}$ for a dual pair $(\mathcal{X}, \mathcal{Y})$ of vector spaces.
- $\mu_n \Rightarrow \mu$, weak convergence of probability measures for a sequence $(\mu_n)_n \subset \mathscr{P}(\mathbf{X})$
- $\nu \ll \mu$, ν is absolutely continuous with respect to μ (for measures)
- $\nu \perp \mu$, measures ν and μ are mutually singular
- $\uparrow$, monotone convergence for non decreasing sequences.
- $\downarrow$, monotone convergence for non increasing sequences.

- SOS, sum of squares
- LP, linear programming
- SDP, semidefinite programming
- GMP, Generalized Moment Problem
- SDr, semidefinite representation (or semidefinite representable)

Index